The mechanics of erosion

M.A.Carson

Monographs in spatial and environmental systems analysis

Series editors R.J.Chorley and D.W.Harvey

1 **Entropy in urban and regional modelling** A.G.Wilson

2 **An introduction to spectral analysis** J.N.Rayner

3 **Perspectives on resource management** T.O'Riordan

4 **The mechanics of erosion** M.A.Carson

Pion Limited, 207 Brondesbury Park, London NW2 5JN

The mechanics of erosion

M.A.Carson

 Pion Limited, 207 Brondesbury Park, London NW2 5JN

Library edition SBN 85086 029 6
Student edition SBN 85086 030 X

Set on IBM 72 Composers by Pion Limited, London
Printed in Great Britain by J.W.Arrowsmith Limited, Bristol.

,00

Preface

Advanced textbooks on particular aspects of geomorphology are appearing at an increasingly rapid rate; many of them incorporate complex mathematical treatments of the mechanics of particular erosional processes. An early example is Bagnold's classic *The Physics of Blown Sand and Desert Dunes*; a similar study of stream erosional processes was afforded by Leopold, Wolman, and Miller's *Fluvial Processes in Geomorphology*; and, more recently, Paterson's *The Physics of Glaciers* has provided a starting-point for investigating the mechanics of glacial erosion. In addition, graduate students in geomorphology are increasingly finding it necessary to delve into the standard textbooks of fluid and solid mechanics, and to extract from them material pertinent to modern process studies in geomorphology.

As far as I am aware, there is no one book which provides a unified introduction to the mechanics of erosional processes and which is aimed at undergraduates in Earth science. Scheidegger's *Theoretical Geomorphology* attempts this task at a more advanced level; however, it assumes a background in mathematics and natural sciences which is not always at the level of the average student in undergraduate Earth science. In writing *The Mechanics of Erosion* I have attempted to unite some of the more elementary fundamental concepts underlying theoretical geomorphology and to provide a full development of these ideas from basic principles. For this purpose I have made extensive use of appendices which in each chapter provide derivations of points in the main text. The monograph is a development of a final-year undergraduate course in geography taught at McGill University.

At the outset, I should warn readers that my use of the term *erosion* in this monograph departs somewhat from the commonly-employed usage of the term. For instance, the American Geological Institute defines erosion as:

> "The group of processes whereby earthy or rock material is loosened or dissolved and removed from any part of the Earth's surface. It includes the processes of weathering, solution, corrasion, and transportation."

In this monograph I shall primarily emphasize the last of these four processes. Solution is undoubtedly an important mechanism of debris removal, but it would be out of place in a treatment of mechanical processes. Much attention has been given, at a superficial level, to corrasion in the previous literature, but I am unable to accept this as a major general mechanism of erosion. Landslides may corrade static material on the failure surface, but the amount of debris corraded is negligible in comparison with the amount of debris already lost through transportation, that is the landslide itself. Introductory textbooks in geomorphology show spectacular illustrations of potholes and striations as examples of fluvial and glacial corrasion respectively, but one must question the quantitative importance of such processes in the degradation of the landscape. Moreover, again, the processes of potholing and glacial abrasion demand that a significant amount of debris is already being transported before corrasion can occur. Accordingly, the key problem is the mechanism by which this loose debris becomes incorporated into the moving fluid, or ice mass, rather than the mechanism by which corrasion takes place.

Admittedly, corrasion cannot be dismissed entirely; the production of rock flour is, for instance, a significant process in glaciated areas. Notwithstanding the attention paid to corrasion by geomorphologists, however, very little is known about the mechanics of this process. Yatsu's *Rock Control in Geomorphology* is one of the few books on Earth science which treats this topic quantitatively. On the other hand, while the mechanics of transportation processes, such as landslides and saltation of debris by streams and wind, have been explored in detail in engineering, it is only recently that this information has been applied to geomorphology. One of the purposes of this monograph is therefore to draw together geomorphically relevant information from the engineering and related sciences into a single introductory text for Earth science students.

Finally, the reader will notice that there is no direct discussion of marine erosional processes. The reason is that this particular form of erosion is highly influenced by the mechanics of wave motion. While this topic is no more complicated than other branches of mechanics that I have introduced in the text, it is sufficiently different that, in order to develop it in the detail accorded to these other branches, I would have needed to expand the text considerably. In addition, a certain amount of unity in the approach would have been lost, and this would have defeated one of the major aims of the monograph. The reader should bear in mind, however, that non-marine processes, such as the slumping of coastal cliffs, form a large part of the erosional activity in coastal areas, and these topics are discussed in various parts of the text.

In preparing the manuscript I have received assistance from many people and this is gratefully acknowledged. In particular, Peter Holland, Eiju Yatsu, and Raymond Yong have provided very thorough criticisms of the whole text. Part of the script was written during my stay in Tucson, Arizona and I wish to acknowledge the hospitality of the Department of Geography, University of Arizona, extended to me at that time. The final script was typed by Susan Ford and Lee Carter, and proofread by my wife Betsy. The final responsibility for errors, of course, remains my own.

M.A.Carson
Assistant Professor of Geography, McGill University, Montréal

Acknowledgements

The author wishes to give credit to the authors listed below, and to others whose work is acknowledged in the appropriate places, for material used in the preparation of some of the figures (all drawn afresh) and tables printed in this book.

R. A. Bagnold, 1953, *The Physics of Blown Sand and Desert Dunes* (Methuen, London): Figures 2.5, 2.10, and A2.4.

H. Carol, 1947, *Journal of Glaciology**: Figure 5.8.

W. S. Chepil, 1961, *Proceedings of Soil Science Society of America:* Table 2.3.

B. R. Colby and C. H. Hembree, 1955, *U.S. Geological Survey*: Figure 2.17.

W. W. Emmett, 1970, *U.S. Geological Survey*: Table 2.4.

J. W. Glen, 1952, *Journal of Glaciology**: Figure 3.5a.

B. L. Hansen and C. C. Langway, 1966, *Antarctic Journal of the United States*: Figure 5.1.

E. W. Lane, E. J. Carlson, and O. S. Hanson, 1949, *Civil Engineering-ASCE***: Figure 2.18b.

L. Lliboutry, 1968, *Journal of Glaciology**: Figures 5.4, A5.2, and A5.3.

R. A. Lohnes and R. L. Handy, 1968, *Journal of Geology* (Chicago University Press, Chicago, Ill.): Figure 4.5.

J. G. McCall, 1960, *Norwegian Cirque Glaciers* (Royal Geographical Society, London): Figures 5.7 and 5.15.

M. Mellor and J. H. Smith, 1966, Cold Regions Research and Engineering Laboratory, Hanover, New Hampshire: Figure 5.11.

A. Mendelson, 1968, *Plasticity: Theory and Application* (MacMillan, New York): Figure A1.1.

F. Müller and A. Iken, 1969, International Association of Scientific Hydrology: Figure 5.5.

J. F. Nye, 1959, *Journal of Glaciology**: Figure 5.2.

R. F. Scott, 1963, *Principles of Soil Mechanics* (Addison-Wesley, Reading, Mass.): Figure 4.9.

W. O. Sellers, 1965, *Physical Climatology* (University of Chicago Press, Chicago, Ill.): Table 2.2.

A. N. Strahler, 1950, *American Journal of Science*: Figure 4.8.

D. W. Taylor, 1948, *Fundamentals of Soil Mechanics* (John Wiley & Sons, New York): Figure 3.5b.

J. Weertman, 1957, 1961, *Journal of Glaciology**: Figures 5.3a and 5.9.

C. M. White, 1940, *Proceedings of the Royal Society (London)*: Figure 2.8.

* by permission of the Glaciological Society

** by permission of the American Society of Civil Engineers

To my parents; and to Betsy

Contents

5 Mechanics of glacial erosion

List of Figures

List of Tables

Systems of units

Three systems of units are used in this monograph: the metric engineering system, the Système International (SI) and the British (and American) engineering system. Comparable units for those dimensions most commonly used in the text are given below.

	Metric engineering system of units	Système International d'unites	British engineering system of units
	kg is weight	kg is mass	lb is weight
Mass		1 kg	0·06852 slugs
Density		1 g cm^{-3}	1·940 slugs ft^{-3}
Unit weight	1000 kg m^{-3}	9807 N m^{-3}	62·42 lb ft^{-3}
Force	1·02 kg	10 N	2·248 lb
Pressure	1·02 kg cm^{-2}	1 bar	14·5 lb in^{-2} (psi)

Notes

1 slug = 1 lb s^2 ft^{-1}
1 newton (N) = 10^5 dynes
1 dyne = 1 g cm s^{-2}
1 bar = 10 N cm^{-2}

Although the Système International is now being rapidly adopted by the scientific community, most of the existing literature in solid and fluid mechanics uses an engineering system of units. This distinction is important: in engineering systems lb and kg are units of *weight*, whereas in SI units kg refers to *mass*. In order to facilitate reference to established texts in solid and fluid mechanics, engineering units form the basis of this book. In general, the metric engineering system has been used (with the British–American equivalent in parenthesis) in the hope that this will assist in the transition to SI units. No special distinction between the three systems is used in the text; it is assumed that the appropriate system is self-evident from the context.

Special mention must be made of the units of angular measurement. In mechanics angles are usually measured in radians, whereas in geomorphology angular values in degrees are more common. Both units are used in this text. The reader should bear in mind that the use of trigonometric functions in calculus assumes that angles are expressed in radians.

[A radian is the value of an angle subtended at the centre of the circle by an arc of length equal to the radius. Thus 2π radians = 360°, $\pi/2$ radians = 90°, 1 radian ≈ 57·3°. It will be seen that for small angles $\tan\theta \approx \sin\theta \approx \theta$, provided θ is in radians.]

1

The concept of stress

Glossary of symbols

The number in square brackets following the definition is the page number where the symbol is first introduced in the text.

b	angle between an arbitrary plane and major principal plane [5]
l, m, n	direction cosines corresponding to $\theta_1, \theta_2, \theta_3$ [8]
s	oblique stress (with both shear and normal components) [11]
x, y, z	Cartesian distance coordinates (z vertical) [2]
δW	weight of small volume [14]
θ	angle between an arbitrary plane and major principal plane [6]
$\theta_1, \theta_2, \theta_3$	angles between a line normal to an arbitrary plane and the directions of the principal stresses [7]
σ	normal stress [2]
$\sigma_1, \sigma_2, \sigma_3$	major, intermediate, and minor principal stresses [2]
σ_m	mean normal stress [9]
σ_{oct}	octahedral normal stress [9]
τ	shear stress [2]
τ_{oct}	octahedral shear stress [9]

1.1 The concept of stress

The term 'stress' represents one of the most fundamental concepts in the mechanics of erosional processes. All forms of erosion are ultimately due to a combination of applied forces acting at, and within, the Earth's surface, and, associated with them, different types of stress.

Stress is often defined simply as the force acting upon a surface *per unit area* of that surface. For instance, if a block of granite weighing 1000 kg (2205 lb) rests upon a horizontal rock platform over an area of 2 m^2 (21·5 ft^2), it could be said that the block exerts a vertical *stress* of 500 kg m^{-2} (0·71 lb in^{-2} or 0·71 psi) on the platform. It could also be said that the block exerts a vertical *pressure* of 500 kg m^{-2} on the platform. Indeed, in this usage, the terms pressure and stress are virtually interchangeable, provided that the direction of the force under consideration is normal to the surface upon which it acts.

Although the illustration above conveys the general meaning of stress quite well—that is, force per unit area—it is misleading in one respect. Pressure and normal stress are not, in fact, synonymous. A more accurate description of the situation considered above would be as follows. The granite block exerts a vertical pressure on the rock platform; in response to this external force (per unit area), internal forces are set up within the platform in order to resist the weight of the block. It is the *internal* force (per unit area) which should be described as a stress. In applied mechanics this distinction is quite generally recognized, but in the literature of geomorphology it is often blurred. The student who finds it difficult to accept the concept of stress may approach it via the more familiar notion of force, provided that these points are borne in mind.

In order to develop additional concepts related to stresses inside a mass of some substance, such as ice or clay, consider the situation in Figure 1.1a, depicting the stresses *acting at a point* (considered as a cubical element for convenience) at depth inside that mass. On each of the six faces, there occur three stresses. The one denoted by σ (sigma) represents the sum of all forces (per unit area) resolved normal to the face and is therefore called the *normal* stress. The other two, denoted by τ (tau), represent the sum of all stresses acting along the face; they are resolved into two directions perpendicular to each other. These stresses are called *shear* stresses or, more rarely, tangential stresses. The notation of the shear stresses is as follows: the first subscript defines the axis normal to the plane in which the shear stress acts; the second subscript defines the direction of the stress within that plane. For every stress acting on a face of the cube, the cube provides a stress, opposite in direction but equal in magnitude, on the material adjacent to that face. In the case of the normal stress σ_z acting on the top face there is, at equilibrium, an equal stress acting upwards on the underneath side of the cube of material above the cube shown in Figure 1.1a. If we inserted something between these two faces, the net effect of these two normal stresses would be to compress that object. Now consider the shear stress τ_{zy}. The reactionary stress, exerted by the cube in the figure on the underneath side of the cube above, is in the opposite direction to τ_{zy}. If we put an object between the two shear stresses, the effect of τ_{zy} would be to try to split, or shear, the object along a surface parallel to the interface between the cubes. The term 'shear stress' is therefore an appropriate name for this type of stress.

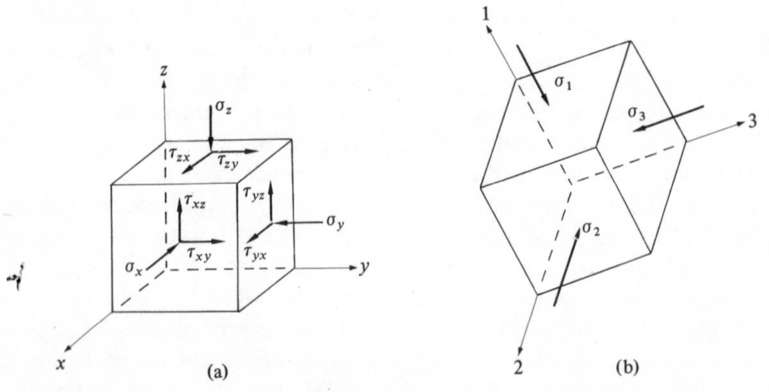

Figure 1.1. Stresses on a cubical element: (a) arbitrary orientation; (b) orientation normal to principal axes.

The orientation of the cube shown in Figure 1.1a is completely arbitrary. We could choose any of an infinite number of orientations to represent the stress conditions at that point. For each orientation there would be, on each face, one normal and two shear stresses; the magnitude of these stresses on any face would vary with the orientation of the cube relative to the external forces acting upon it. Now we arrive at a very important point. At *one* particular orientation, all the shear stresses acting on the cube become zero, and only normal stresses remain (this is shown in Figure 1.1b). In other words, at any point in a mass there exist three mutually perpendicular surfaces on which only normal stresses act. These surfaces are called the *principal planes* at that point; the normal stresses acting on these planes are termed *principal stresses*, and the directions of the principal stresses are referred to as *principal stress axes.*

An actual example may make this point clearer. The strength of a clay soil is often determined by subjecting a cylindrical specimen to compression in an apparatus termed a triaxial cell. In this apparatus (Figure 1.2a) the specimen, which is enclosed in a very thin rubber membrane, is surrounded by water at a certain pressure. This pressure acts along the normal to the specimen at all points and is equivalent, for the cylindrical specimen, to σ_x and σ_y on the cube shown in Figure 1.1a. The difference between Figure 1.1a (a general model) and Figure 1.2a (a particular case) is that $\sigma_x = \sigma_y$ for the conventional triaxial cell, because the water pressure is the same on all sides of the specimen.

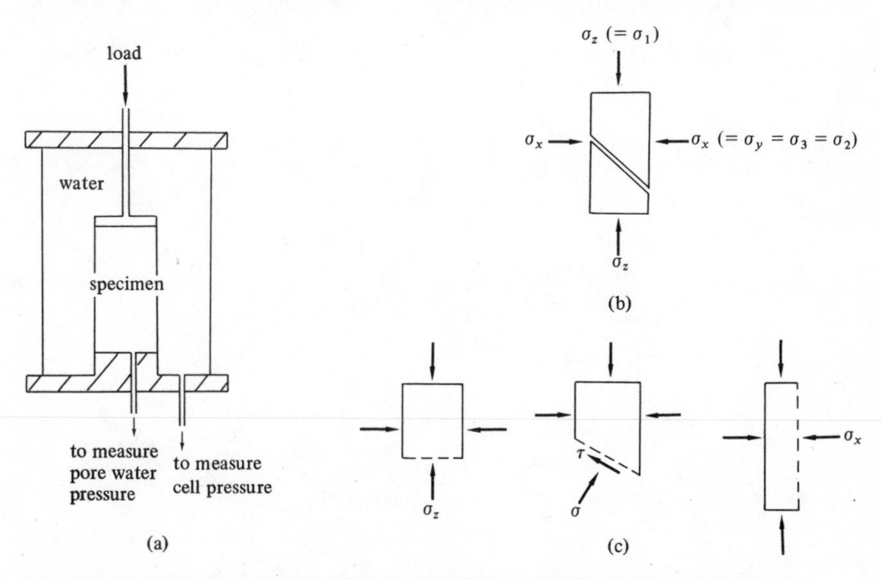

Figure 1.2. The conventional triaxial test: (a) apparatus; (b) stress conditions; (c) stresses on various sections in the specimen.

Once the specimen is set up in the triaxial cell in this way, it is subjected to an increasing principal stress in the vertical direction (σ_z). Eventually this vertical stress becomes so large that the specimen will shear (Figure 1.2b), and it does so along a surface inclined at an angle intermediate between the vertical axis and the horizontal plane. Why does the specimen fail by shearing along this plane rather than along a horizontal or a vertical plane? The reason is as follows. There are no *shear* stresses on planes aligned vertically or horizontally. If, however, we consider any plane between the horizontal plane and the vertical direction, and resolve stresses (or forces) along that plane (Figure 1.2c), there is now a shear stress acting on the surface. In this case, then, the horizontal plane and any two perpendicular vertical planes are principal planes. The three principal stresses are ranked in order of magnitude: major (σ_1), intermediate (σ_2), and minor (σ_3). In the triaxial test, the vertical stress is the major principal stress. The intermediate and minor principal stresses are, in the conventional triaxial test, identical and equal to the all-round cell pressure.

Sometimes we may be confronted with a problem in which we know the intensity of the stresses on the principal planes, and we wish to know the magnitude of the shear and normal stresses acting on another plane inclined in a particular way to the principal planes. The solution for this is quite simple, but before dealing with the general three-dimensional problem (Figure 1.3a), it is useful to consider the special two-dimensional case (Figure 1.3b) in which the plane is *not* tilted toward the σ_2 axis. This version is commonly used in the analysis of very wide landslides and it will be employed in Chapter 4. The solution for the two-dimensional

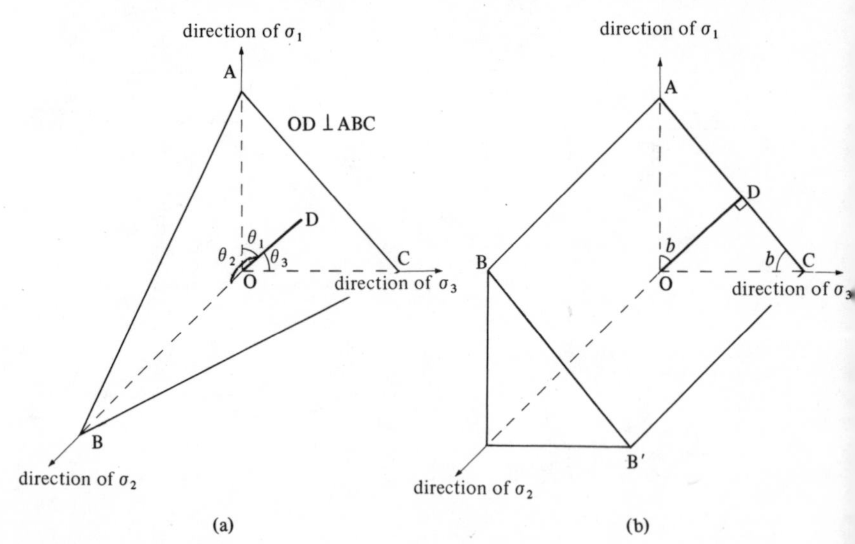

Figure 1.3. Orientation of a plane relative to principal axes: (a) general case; (b) two-dimensional case.

case may be obtained algebraically or geometrically. We shall first of all consider the algebraic treatment. Consider the stresses acting on a plane that makes an angle b with the major principal plane. In particular, consider a prism-shaped element of material, as in Figure 1.4, bounded by surfaces provided by the major principal plane, the minor principal plane, and the plane under study. The problem is to determine τ and σ in relation to σ_1, σ_3, and b. If the prism is assumed to be in equilibrium and we resolve forces perpendicular to, and afterwards along, the plane AB, we obtain:

$$\begin{aligned}\sigma\,\delta s &= \sigma_3\,\delta z\cos(\pi/2-b)+\sigma_1\,\delta x\cos b\\ &= \sigma_3\,\delta z\sin b+\sigma_1\,\delta x\cos b\;;\end{aligned} \tag{1.1}$$

and

$$\begin{aligned}\tau\,\delta s &= \sigma_1\,\delta x\cos(\pi/2-b)-\sigma_3\delta z\cos b\\ &= \sigma_1\,\delta x\sin b-\sigma_3\delta z\cos b\;.\end{aligned} \tag{1.2}$$

Now we know that $\delta z/\delta s = \sin b$ and $\delta x/\delta s = \cos b$; if, therefore, we substitute in Equations (1.1) and (1.2) for δs, we obtain

$$\sigma = \sigma_3\sin^2 b+\sigma_1\cos^2 b \tag{1.3}$$

and

$$\begin{aligned}\tau &= \sigma_1\sin b\cos b-\sigma_3\sin b\cos b\\ &= (\sigma_1-\sigma_3)\sin b\cos b\;.\end{aligned} \tag{1.4}$$

Usually these two equations are expressed in double-angle form. If, in Equation (1.3), we substitute $\cos^2 b = \frac{1}{2}(1+\cos 2b)$ and $\sin^2 b = \frac{1}{2}(1-\cos 2b)$, we obtain

$$\begin{aligned}\sigma &= \sigma_3(\tfrac{1}{2}-\tfrac{1}{2}\cos 2b)+\sigma_1(\tfrac{1}{2}+\tfrac{1}{2}\cos 2b)\\ &= \tfrac{1}{2}(\sigma_1+\sigma_3)+\tfrac{1}{2}(\sigma_1-\sigma_3)\cos 2b\;.\end{aligned} \tag{1.5}$$

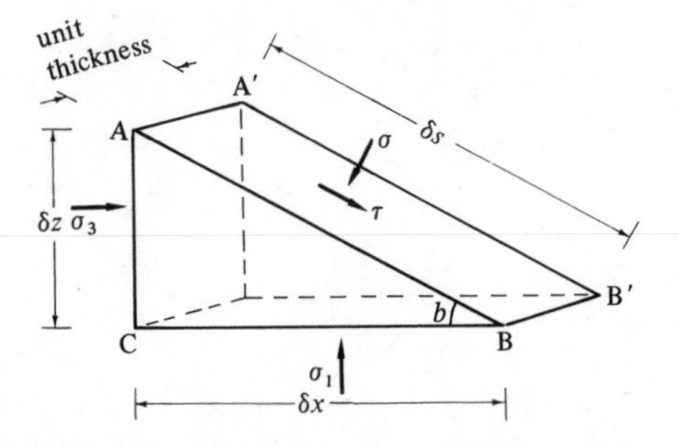

Figure 1.4. Stresses on a wedge-shaped element: two-dimensional case.

Similarly, Equation (1.4) reduces to

$$\tau = \tfrac{1}{2}(\sigma_1 - \sigma_3)\sin 2b \ . \qquad (1.6$$

These equations are important. They tell us the magnitude of the shear stress and normal stress on *any* surface in a two-dimensional situation provided that we know the magnitudes and directions of the principal stresses and the angle that the section makes with them. We shall encounter the equations in sections of both Chapters 4 and 5.

Frequently Equations (1.5) and (1.6) are depicted graphically by means of *Mohr's circle of stress* (Figure 1.5). In this diagram, which represents the state of stress on *different sections through a single point* in a mass, the normal stresses are plotted on the horizontal axis and the shear stress on the vertical axis. The distance OA is drawn equal to σ_3 and denotes the stress on a section parallel to the minor principal plane ($\tau = 0$, $\sigma = \sigma_3$); OB is drawn equal to σ_1 and represents the stress on a section parallel to the major principal plane ($\tau = 0$, $\sigma = \sigma_1$). Consider the point C. The normal stress on the section represented by this point is given by OA plus AD; the shear stress is equal to CD. It is fairly simple to show that this state of stress corresponds to that for a plane making an angle θ (theta) (θ being defined as in Figure 1.5) with the major principal plane. From the diagram,

$$\mathrm{CD} = \mathrm{CX} \times \sin 2\theta$$

and, with CD = τ, and CX = AX = $\tfrac{1}{2}(\sigma_1 - \sigma_3)$, we have

$$\tau = \tfrac{1}{2}(\sigma_1 - \sigma_3)\sin 2\theta \ .$$

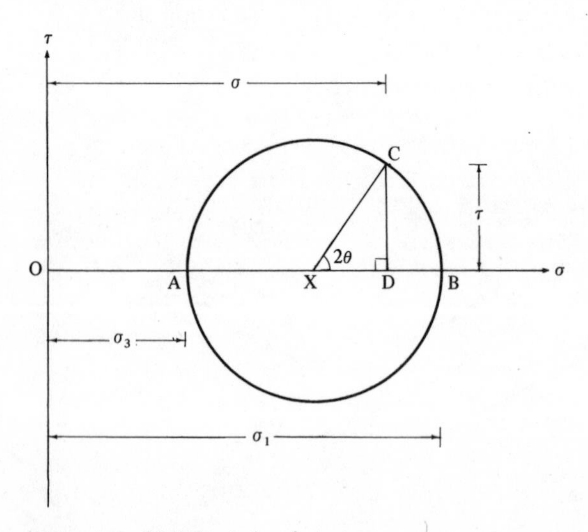

Figure 1.5. Mohr's circle of stress: $\sigma_1 = \sigma_2$ or $\sigma_2 = \sigma_3$.

By comparison with Equation (1.6), we can see that this is the value of the shear stress on a plane making an angle θ with the major principal plane. Let us now consider the normal stress on the plane represented by point C. From the diagram,

$$AD = AX + XD = AX + CX\cos 2\theta$$

and, with

$$CX = AX = \tfrac{1}{2}(\sigma_1 - \sigma_3)\,,$$

we have

$$AD = \tfrac{1}{2}(\sigma_1 - \sigma_3)(1 + \cos 2\theta)\,.$$

Now,

$$\begin{aligned}\sigma &= OD = OA + AD \\ &= \sigma_3 + \tfrac{1}{2}(\sigma_1 - \sigma_3) + \tfrac{1}{2}(\sigma_1 - \sigma_3)\cos 2\theta\end{aligned}$$

so that

$$\sigma = \tfrac{1}{2}(\sigma_1 + \sigma_3) + \tfrac{1}{2}(\sigma_1 - \sigma_3)\cos 2\theta\,.$$

By comparison with Equation (1.5) it can be seen that this is the value of the normal stress acting on a plane making an angle θ with the major principal plane. We thus arrive at a very important general conclusion. Any point on the circle in Figure 1.5 depicts the state of stress (that is, the values of the shear and normal stresses) for a particular section passing through a point with known major and minor principal stress values. The actual stresses acting on any section making an angle θ ($=b$) with the major principal plane (point B: $\sigma = \sigma_1$, and $\tau = 0$) can thus be obtained by locating the point C on the Mohr circle; this is done by simply drawing in a radius at an angle 2θ to the σ axis. In Section 3.2, we shall see how the Mohr circle of stress, together with knowledge about the strength of the mass, can be used to predict the orientation of failure surfaces.

Although the two-dimensional approach has been widely and successfully used in field studies, it is not entirely satisfactory to students of theoretical soil mechanics because it allows no consideration to be made for different values of the intermediate principal stress. Moreover, as noted in Chapter 5 dealing with glacier flow, not all field problems are simple two-dimensional ones. Brief reference to the general case (Figure 1.3a) must, therefore, be made here. The face ABC forms part of a plane inclined towards all three principal axes. The orientation of this face can be defined in many ways. One method would be in terms of the angle between each principal axis and a line from the origin normal to the face. If θ_1, θ_2, θ_3 are defined as in the diagram, it is evident that θ_1 is comparable to b in the two-dimensional case. Conventionally, in texts on mechanics, the orientation of the face ABC is defined in terms of the cosines of these angles, rather than the angles themselves:

$$l = \cos\theta_1, \qquad m = \cos\theta_2, \qquad n = \cos\theta_3,$$

and, again, for the two-dimensional case, $l = \cos b$, $m = \cos(\pi/2) = 0$, and $n = \cos(\pi/2 - b) = \sin b$.

From a consideration of the forces acting on the tetrahedron OABC, it can be shown (Appendix 1.1) that the stresses on the face ABC are given by

$$\sigma = l^2\sigma_1 + m^2\sigma_2 + n^2\sigma_3 \tag{1.7}$$

and

$$\tau^2 = l^2\sigma_1^2 + m^2\sigma_2^2 + n^2\sigma_3^2 - \sigma^2, \tag{1.8}$$

where σ is the stress normal to ABC and τ is the shear stress acting in the direction of maximum shear; τ could, of course, be resolved into any two components acting normal to each other within the ABC face. Note that, for the two-dimensional case ($m = 0$), Equations (1.7) and (1.8) reduce to Equations (1.3) and (1.4).

It is possible to depict the triaxial state of stress graphically just as in the two-dimensional case. This is shown in Figure 1.6. The values of σ and τ on any plane are represented by a particular point located within the shaded area. (The development of this is provided in Appendix 1.2.) Note that the state of stress is no longer confined to the outer circle, as in the two-dimensional case. The reason is that the stress values are now

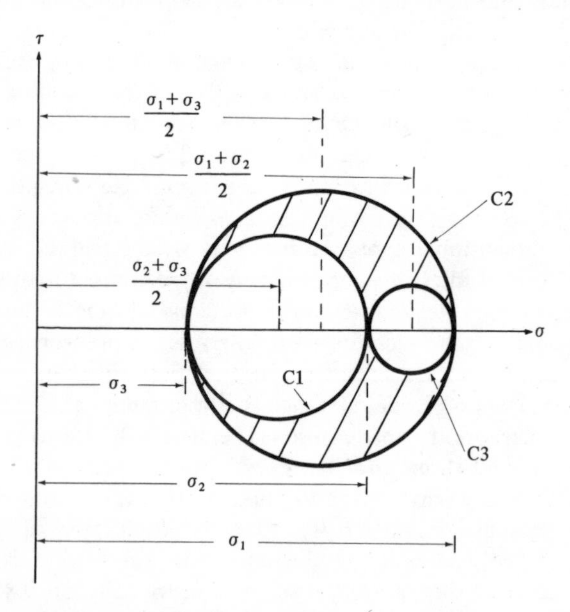

Figure 1.6. Mohr's stress diagram: $\sigma_1 > \sigma_2 > \sigma_3$.

influenced by the intermediate, as well as the major and minor, principal stress. Note, further, that, if two of the principal stresses are equal ($\sigma_1 = \sigma_2$ or $\sigma_2 = \sigma_3$), under these *particular* conditions the triaxial state of stress reduces to the two-dimensional case. This may be visualized with reference to Figure 1.6; if σ_2 is equal to either of the other principal stresses, one of the inside circles diminishes to a point and the other enlarges until it is coincident with the outer circle. It can be shown algebraically (Appendix 1.3), that Equations (1.7) and (1.8) reduce to Equations (1.3) and (1.4) if $\sigma_2 = \sigma_1$ or if $\sigma_2 = \sigma_3$. The two-dimensional equations can, therefore, be used in the computation of σ and τ on shear surfaces in conventional triaxial tests in which, as noted previously, $\sigma_2 = \sigma_3$.

The face ABC in Figure 1.3a is inclined at arbitrary angles to the three axes. At certain orientations the plane may assume very important properties. One such position is given by $l = m = n = 1/\sqrt{3}$ (AB = BC = CA); this plane is called the *octahedral* plane. The reason for this name is that construction of faces similar to ABC in the other seven quadrants (three above the σ_2-σ_3 plane, and four below) would produce a regular octahedron. The stresses on this plane (σ_{oct} normal to it, and τ_{oct} along it in the direction of maximum shear) are obtained by inserting $l = m = n = 1/\sqrt{3}$ in Equations (1.7) and (1.8):

$$\sigma_{oct} = \tfrac{1}{3}(\sigma_1+\sigma_2+\sigma_3) = \sigma_m\,, \tag{1.9}$$

$$\tau_{oct}^2 = \tfrac{1}{3}[(\sigma_1-\sigma_m)^2+(\sigma_2-\sigma_m)^2+(\sigma_3-\sigma_m)^2]\,, \tag{1.10}$$

or

$$\tau_{oct}^2 = \tfrac{1}{9}[(\sigma_1-\sigma_2)^2+(\sigma_2-\sigma_3)^2+(\sigma_3-\sigma_1)^2]\,, \tag{1.11}$$

where σ_m is the mean normal stress. (The octahedral normal and shear stresses are given in terms of general orthogonal stresses, σ_x, σ_y, and σ_z, in Appendix 3.3.) The concept of the octahedral shear stress will be used in Chapter 5 on glacier flow.

Finally we should note that special circumstances prevail in the case of fluids. Simple experiments can be undertaken to show that for fluids in a *static* condition (fluid *flow* is discussed in Chapter 2): (1) there are no *shear* stresses on *any* section through a point; and (2) the normal stress at any point is the same on *all* sections passing through that point. It is for this reason that it is meaningful to speak of fluid pressure; in a static fluid, pressure is the same in all directions at a point. (As we shall see in Chapter 3, this is not the case for solids.) These two properties of a static fluid mean that the state of stress is represented by a point on the σ axis in the Mohr diagram of Figure 1.5. Finally, note (Appendix 1.4) that the second property listed above follows directly from the first.

Bibliography

Hill, R., 1950, *The Mathematical Theory of Plasticity* (Clarendon Press, Oxford).
Hoffman, O., Sachs, G., 1953, *Introduction to the Theory of Plasticity for Engineers* (McGraw-Hill, New York).
Jaeger, J. C., 1962, *Elasticity, Fracture and Flow* (Methuen, London).
Malvern, L. E., 1969, *Introduction to the Mechanics of a Continuous Medium* (Prentice-Hall, Englewood Cliffs, N.J.).
Mendelson, A., 1968, *Plasticity: Theory and Application* (MacMillan, New York).

Appendix 1

1.1 Proof that the stresses on any plane oriented relative to the principal axes with direction cosines of l, m, and n are given by

$$\sigma = l^2\sigma_1 + m^2\sigma_2 + n^2\sigma_3 \qquad \text{[Equation (1.7)]},$$

$$\tau^2 = l^2\sigma_1^2 + m^2\sigma_2^2 + n^2\sigma_3^2 - \sigma^2 \qquad \text{[Equation (1.8)]}.$$

In Figure A1.1, tensile stresses are taken as positive and compressive stresses as negative (unlike Figure 1.1) in order to conform with conventional treatments in mechanics. Resolving forces in the x direction, at equilibrium, we have:

$$(\text{ABC})s_x = (\text{AOB})\sigma_x + (\text{AOC})\tau_{yx} + (\text{BOC})\tau_{zx}, \tag{A1.1}$$

where the parentheses denote the area of the individual faces. From the geometry of the element OABC, it is easily shown that AOB $= l\,\delta A$, AOC $= m\,\delta A$, and BOC $= n\,\delta A$, where δA denotes the area of ABC. Accordingly, Equation (A1.1) becomes

$$s_x = l\,\sigma_x + m\,\tau_{yx} + n\,\tau_{zx} \tag{A1.2}$$

and, similarly,

$$s_y = l\,\tau_{xy} + m\,\sigma_y + n\,\tau_{zy} \tag{A1.3}$$

$$s_z = l\,\tau_{xz} + m\,\tau_{yz} + n\,\sigma_z\,. \tag{A1.4}$$

Projecting s_x, s_y, s_z onto ON, we obtain

$$\sigma = l\,s_x + m\,s_y + n\,s_z\,,$$

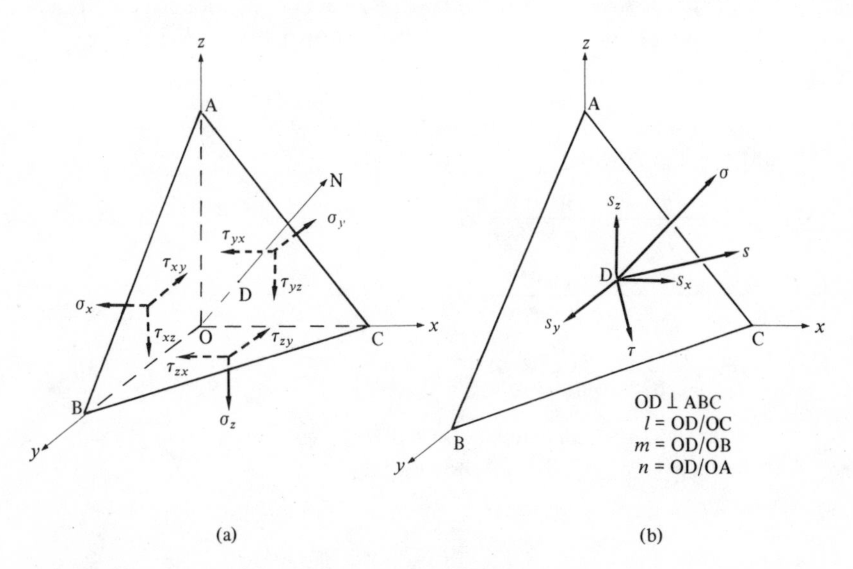

Figure A1.1. Stresses on a right tetrahedral element: (a) stresses on the orthogonal faces; (b) stresses on the hypotenuse face (after Mendelson, 1968).

or

$$\sigma = (l^2\sigma_x + l\,m\,\tau_{yx} + l\,n\,\tau_{zx}) + (m\,l\,\tau_{xy} + m^2\sigma_y + m\,n\,\tau_{zy}) + (n\,l\,\tau_{xz} + n\,m\,\tau_{yz} + n^2\sigma_z)\,.$$

It can be shown (Mendelson, 1968, p.29) that, for solids, $\tau_{yx} = \tau_{xy}$, $\tau_{zy} = \tau_{yz}$, and $\tau_{zx} = \tau_{xz}$. The equation for the normal stress is thus reduced to

$$\sigma = l^2\sigma_x + m^2\sigma_y + n^2\sigma_z + 2(l\,m\tau_{xy} + m\,n\tau_{yz} + n\,l\tau_{zx})\,. \qquad \text{(A1.5)}$$

From elementary mechanics,

$$\tau^2 = s^2 - \sigma^2$$

so that

$$\tau^2 = (s_x^2 + s_y^2 + s_z^2) - \sigma^2 \qquad \text{(A1.6)}$$

in which substitution for s_x, s_y, and s_z would yield an equation for the shear stress comparable to Equation (A1.5).

If we now take the coordinate axes in the principal directions ($x = 1$, $y = 2$, $z = 3$), the equations simplify because all the shear stresses are zero; from Equations (A1.2) to (A1.6), we obtain Equations (1.7) and (1.8):

$$\sigma = l^2\sigma_1 + m^2\sigma_2 + n^2\sigma_3\,,$$

$$\tau^2 = l^2\sigma_1^2 + m^2\sigma_2^2 + n^2\sigma_3^2 - \sigma^2\,.$$

1.2 The development of Mohr's diagram for the three-dimensional case (Figure 1.6)

From elementary geometry, it can be shown that

$$l^2 + m^2 + n^2 = 1\,; \qquad \text{(A1.7)}$$

Equations (A1.7), (1.7), and (1.8) may now be solved for values of l, m, and n. This yields:

$$l^2 = \frac{\tau^2 + (\sigma - \sigma_2)(\sigma - \sigma_3)}{(\sigma_1 - \sigma_2)(\sigma_1 - \sigma_3)}\,, \qquad \text{(A1.8)}$$

$$m^2 = \frac{\tau^2 + (\sigma - \sigma_3)(\sigma - \sigma_1)}{(\sigma_2 - \sigma_3)(\sigma_2 - \sigma_1)}\,, \qquad \text{(A1.9)}$$

$$n^2 = \frac{\tau^2 + (\sigma - \sigma_1)(\sigma - \sigma_2)}{(\sigma_3 - \sigma_1)(\sigma_3 - \sigma_2)}\,. \qquad \text{(A1.10)}$$

Now, $l^2 \geqslant 0$ and, because $\sigma_1 \geqslant \sigma_2 \geqslant \sigma_3$, $(\sigma_2 - \sigma_2)(\sigma_1 - \sigma_3) \geqslant 0$, so that, from Equation (A1.8), it follows that

$$\tau^2 + (\sigma - \sigma_2)(\sigma - \sigma_3) \geqslant 0\,,$$

or

$$\tau^2 + [\sigma - \tfrac{1}{2}(\sigma_2 + \sigma_3)]^2 \geqslant [\tfrac{1}{2}(\sigma_2 - \sigma_3)]^2\,. \qquad \text{(A1.11)}$$

From elementary geometry, an equation of the form

$$(x-h)^2+(y-k)^2 = r^2$$

is the locus of a circle of radius r and centre (h, k). Thus, in the case when (A1.11) represents an equality, it is seen that it describes a circle of radius $\frac{1}{2}(\sigma_2-\sigma_3)$ and centre $[\frac{1}{2}(\sigma_2+\sigma_3), 0]$ relative to (σ, τ) coordinate axes. This is the circle C1 in Figure 1.6. The region defined by (A1.11) therefore lies outside C1 which is a boundary. The state of stress on any plane will, therefore, never be given by values of τ and σ inside C1. Similar examination of (A1.9) and (A1.10) shows that

$$\tau^2+[\sigma-\tfrac{1}{2}(\sigma_1+\sigma_3)]^2 \leqslant [\tfrac{1}{2}(\sigma_1-\sigma_3)]^2 \qquad \text{(A1.12)}$$

and

$$\tau^2+[\sigma-\tfrac{1}{2}(\sigma_1+\sigma_2)]^2 \geqslant [\tfrac{1}{2}(\sigma_1-\sigma_2)]^2 \qquad \text{(A1.13)}$$

which, in terms of Figure 1.6, relate to circles C2 and C3 respectively. Together, (A1.11), (A1.12), and (A1.13) indicate that the point describing the state of stress on any plane must lie in the shaded area of Figure 1.6.

1.3 Proof that Equations (1.7) and (1.8) reduce to Equations (1.3) and (1.4) if two of the principal stresses are equal in magnitude

If $\sigma_2 = \sigma_3$, Equation (1.7) reduces to

$$\sigma = l^2\sigma_1+(m^2+n^2)\sigma_3 \ ;$$

substituting $m^2+n^2 = 1-l^2$ [from (A1.7)], we obtain

$$\sigma = l^2\sigma_1+(1-l^2)\sigma_3 \ ,$$

which is identical to Equation (1.3) with $l^2 = \cos^2 b$ and $(1-l^2) = \sin^2 b$. Siınilarly, Equation (1.8) reduces to

$$\tau^2 = l^2\sigma_1^2+(1-l^2)\sigma_3^2-[l^2\sigma_1+(1-l^2)\sigma_3]^2 \ ,$$

or

$$\tau^2 = l^2(1-l^2)(\sigma_1-\sigma_3)^2 \ ,$$

which is identical to Equation (1.4) with $l = \cos b$.

1.4 Proof that, if there are no shear stresses in a static fluid, the normal stress at any point is the same in all directions

Refer to Figure A1.2 which, for simplicity, shows the two-dimensional case. At equilibrium, resolving forces horizontally yields

$$\sigma_x\,\delta z+\tau\frac{\delta x}{\sin\theta}\sin\theta = \sigma\frac{\delta z}{\cos\theta}\cos\theta+\tau_{zx}\delta x$$

and, with $\tau = \tau_{zx} = 0$, we obtain

$$\sigma_x = \sigma \ .$$

Similarly, resolving forces in the z direction yields

$$\sigma_z \, \delta x = \tau_{xz} \delta z + \sigma \frac{\delta x}{\sin\theta} \sin\theta + \tau \frac{\delta z}{\cos\theta} \cos\theta + \delta W$$

and, with $\tau = \tau_{xz} = 0$,

$$\sigma_z \, \delta x = \sigma \, \delta x + \delta W \, .$$

As δx, $\delta z \to 0$, $\delta W \to 0$ and, therefore,

$$\sigma_z = \sigma \, .$$

The argument is easily extended to the three-dimensional case.

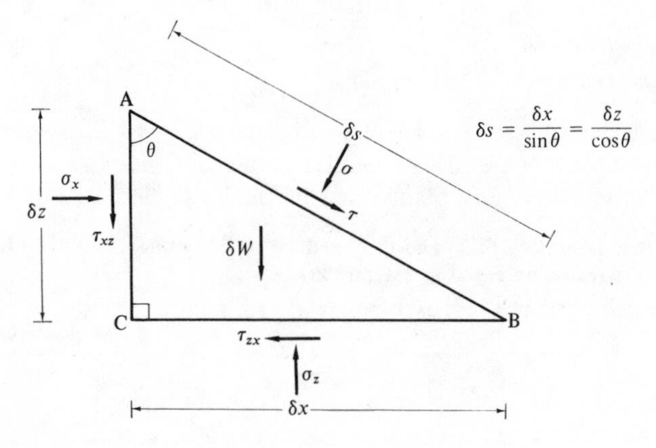

Figure A1.2. Stresses on a two-dimensional element.

Mechanics of fluid erosion

Glossary of symbols

a specified distance above stream bed [43]; surface area of a particle [33]; coefficient [40]
b exponent [40]; velocity gradient in a semi-logarithmic plot [21]
c sediment concentration at a point in fluid [42]; coefficient of proportionality [39]
c_a sediment concentration at height a [43]
d particle diameter [21]
d_c critical particle diameter (largest that can be moved under given flow condition) [55]
f Darcy–Weisbach resistance coefficient [56]
f function of [40]
g acceleration due to gravity [25]
h_f friction head [58]
i slope angle [26]; exponent [43]
k' roughness parameter for mobile boundary [38]
l Prandtl's mixing length [23]
m, m' mass and submerged mass of a particle [27]
n Manning's roughness coefficient [57]; number of particles on a unit area of boundary surface [26]
p rate of loss of momentum per unit length in direction of flow [37]
p_s probability of a grain being dislodged in saltation in unit time [40]
q_b total mass of bed load per unit width of flow per unit time [38]
q_c mass of creeping bed load discharged per unit width per unit time [38]
q_s mass of saltating bed load discharged per unit width per unit time [37]
s slope of water surface (in radians) [25]
t time (duration) [37]
u velocity at a point in direction of flow [21]
$\bar{u}$ mean velocity through a cross-section normal to flow [55]
u_* shear velocity [24]
u_{*_c} critical shear velocity [28]
u_0 ambient velocity [19]
u_t threshold velocity [38]
v vertical velocity [38]
w' submerged weight of a particle ($= m'g$) [40]
w_c competence of stream under given conditions of flow (comparable to d_c) [55]

z height above boundary surface [21]
z_0 roughness of boundary (thickness of layer of stationary fluid) [21]

A area [20]; parameter used by Bagnold, comparable to square root of Shields' entrainment function for flow over flat surface [28]; constant of integration [61]
A_1, A_2 parameters used by Einstein to describe particle shape [39]
B impact coefficient [38]
C Chezy's roughness coefficient [57]; constant of proportionality [33]; constant of integration [24]
C_D drag coefficient [33]
C_L lift coefficient [34]
D measure of length in Reynolds' Number [19]; depth of fluid flow [25]; diameter of pipe [56]
F_D drag force on a single particle [26]
F_L lift force on a single particle [33]
L leap of a saltating particle [37]; length downstream in fluid flow [25]
N number of particles saltating past a line of unit width [37]
N_R Reynolds Number [19]
P_w wetted perimeter [20]
R hydraulic radius [20]
T boundary shear or tractive force [26]
W width of fluid flow [25]

α coefficient of proportionality [38]
γ_w unit weight of water [25]
δ_0 thickness of laminar sublayer [20]
ϵ_m kinematic eddy viscosity [23]
ϵ_s mixing coefficient for sediment [42]
η packing coefficient ($= nd^2$) [27]
κ von Kármán's constant ($= 0{\cdot}4$) [23]
ν kinematic molecular viscosity [19]
ξ parameter relating to graded streams ($= W^{1/3}Rs/d^{1/2}$) [48]
ρ_f mass density of fluid [19]
ρ_s mass density of sediment (particle) [27]
ρ' submerged mass density of a particle ($= \rho_s - \rho_f$) [27]
τ fluid shear stress [22]
τ_0 boundary shear stress [20]; unit tractive force [25]
$\overline{\tau_0}$ average boundary shear stress [25]
τ_c critical tractive force per unit area [27]
ϕ angle of interlock among particles [27]
ω fall velocity of a particle [36]

Φ Einstein's bed load function [40]
Ψ parameter used by Einstein approaching reciprocal of Shields' entrainment function at threshold of bed load movement [40]

2.1 Elementary fluid dynamics

The flow of any fluid within confining boundary surfaces is slowed down by shear forces imposed by those surfaces on the fluid in a direction opposite to the flow. Water flowing in stream channels (Figure 2.1) is, for instance, retarded by shear resistance, or *surface drag*, along the banks, the stream bed and, to a very much smaller extent, along the air-water interface.

Theory and experimental observation indicate that a very thin layer of fluid in contact with the boundary surface is actually slowed down completely so that, relative to the boundary, the velocity of this thin layer is zero. (This is true for all fluids except gases at very low pressure which can slip past the surface.) The shear resistance between two fluid layers is not as effective as that between the boundary surface and the adjacent fluid layer; as a result, fluid flow in layers further from the boundary is only partly slowed down. As indicated in Figure 2.2a the velocity of fluid flow increases with distance away from the boundary. This is commonly observed in nature: wind velocities increase with height above the ground surface; stream channel flow is faster away from the bed and the banks of the channel.

The actual velocity profile, that is, the velocity plotted against the distance from the boundary, depends on the type of fluid flow. The type of flow illustrated in Figure 2.2a, in which layers of fluid glide past each other in parallel directions is called *laminar flow* (see Appendix 2.1 for definitions of common fluid mechanics terms). In this type of flow the velocity distribution is parabolic (Figure 2.2b); the explanation for this is deferred until Section 3.3. In nature, however, fluid flow is usually very erratic, and movement of some fluid transverse to the mean direction of flow takes place (Figure 2.2c); this is described as *turbulent flow*. As indicated in Figure 2.2d, the velocity gradient in this type of flow is much

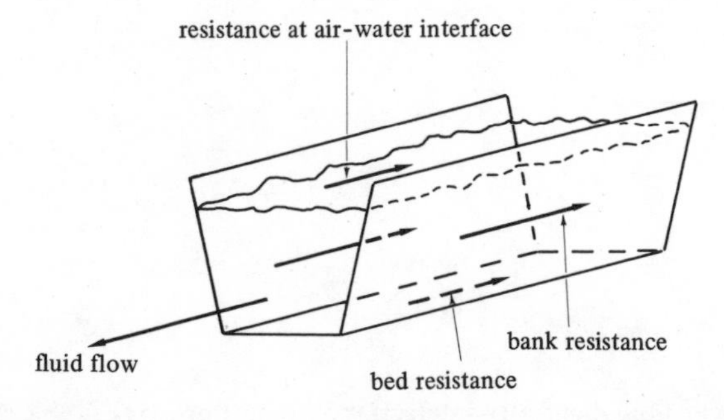

Figure 2.1. Boundary resistance to flow in a stream channel.

greater nearer the boundary, and much smaller away from it, in comparison with laminar flow. The reason is quite simple. The transverse components of flow in turbulent conditions transport slow-moving fluid from near to the boundary towards the centre of flow (decreasing the velocity there) and transport fast-moving fluid from the centre towards the boundary (increasing the velocity there). The question now arises: What

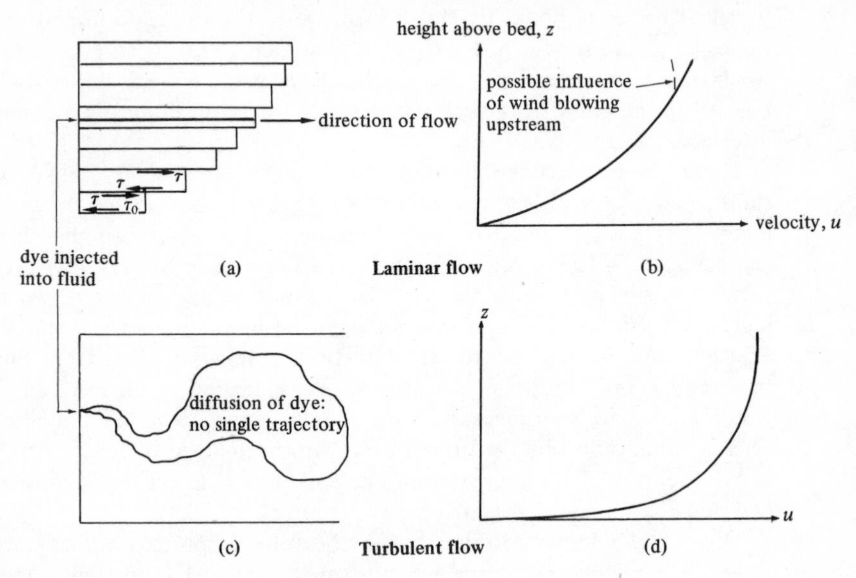

Figure 2.2. Laminar and turbulent flow conditions.

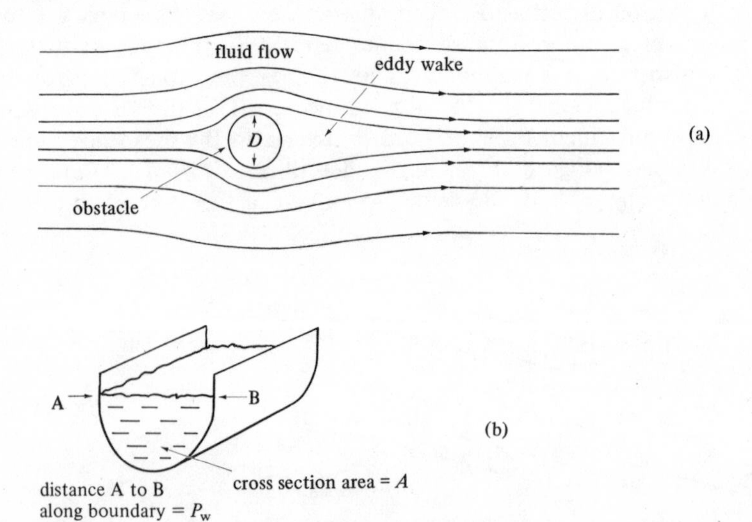

Figure 2.3. Two types of fluid-boundary interaction in fluid flow: (a) flow past an obstacle; (b) flow within a confining surface.

controls whether fluid flow is turbulent or laminar? In order to answer this we need to introduce the dimensionless parameter called the Reynolds Number.

Shapiro (1961) has provided an excellent account of this problem in relation to the amount of turbulence produced by the insertion of an obstacle into the path of fluid flow. Clearly the amount of turbulence produced in the flow *around the object* (Figure 2.3a) depends partly on the shape of the object. For a fixed shape, it also depends on the parameter

$$N_R = \frac{Du_0}{\nu}, \tag{2.1}$$

the Reynolds Number, in which D is a measure of the size of the obstacle in dimensions of length; u_0 is the velocity of the ambient fluid and ν (nu) is the *kinematic molecular viscosity*. (For convenience, the discussion here is couched in terms of kinematic rather than dynamic viscosity; the relationship between the two forms is briefly dealt with in Appendix 2.2.) Kinematic viscosity is essentially a measure of the interference between adjacent layers of fluid during flow; it is measured in dimensions of area/time (L^2T^{-1}). Molecular viscosity is, for any fluid, a function of temperature (Table 2.1), and for compressible fluids also dependent on pressure, but it is not influenced by the type, or velocity, of flow. The degree of turbulence in a fluid-object system is directly related to the Reynolds Number.

Table 2.1. Mass density and kinematic molecular viscosity of air and water at different temperatures and standard atmospheric pressure.

Temperature		Mass density, ρ_f, × 10^3		Kinematic molecular viscosity, ν	
				× 10^5	× 10^2
(°F)	(°C)	(slugs ft^{-3})	(g cm^{-3})	(ft^2 s^{-1})	(cm^2 s^{-1})
Air					
0	−18	2·68	1·38	12·6	11·7
20	− 6·5	2·57	1·32	13·6	12·6
40	4·5	2·47	1·27	14·6	13·6
60	15·5	2·37	1·22	15·8	14·7
80	26·5	2·28	1·18	16·9	15·7
100	38	2·20	1·13	18·0	16·8
Water					
32	0	1940	1000	1·93	1·80
40	4·5	1940	1000	1·66	1·55
50	10	1940	1000	1·41	1·31
60	15·5	1938	999	1·22	1·14
70	21	1936	998	1·06	0·98
80	26·5	1934	997	0·93	0·86

A similar relationship holds for the flow of fluid *through a channel* confined by a boundary surface (Figure 2.3b). The term D in the Reynolds number in this case relates to the dimensions of the channel. It is usually taken as equal to four times the *hydraulic radius,* defined as

$$R = \frac{A}{P_w} , \qquad (2.2)$$

where A is the cross-sectional area of the flow and P_w is the length of the wetted perimeter. It is easily shown that, for wide channels, the hydraulic radius rapidly approaches the depth of flow in value. For open-channel flow, it has been observed (Chow, 1959, p.8) that for $N_R < 2000$ flow is laminar, for $N_R > 10000$ flow is turbulent, and for $2000 < N_R < 10000$ either type of flow may occur. Most channel flow is turbulent. (Some workers use $D = R$ in the Reynolds number; accordingly the limiting values for the transition between laminar flow and turbulent flow are, in this case, $500 < N_R < 2500$.) Overland flow on hillsides may be either laminar or turbulent (Emmett, 1970) depending on the depth of flow. Brunt (1934) calculated (using D as the distance between the ground surface and the tropopause) that winds with velocities greater than 1 m s^{-1} will usually be turbulent.

Although most fluid flow in nature is thus turbulent, laminar conditions exist in a very thin layer, appropriately known as the *laminar sublayer,* adjacent to the boundary surface. Experiments have shown that the thickness of this sublayer, δ_0 (delta), is determined by the flow conditions:

$$\delta_0 = 11 \cdot 6 \nu \sqrt{\frac{\rho_f}{\tau_0}} \qquad (2.3)$$

where ρ_f (rho) is the mass density of the fluid (see Table 2.1) and τ_0 is the shear stress within the fluid at the boundary. We shall see later (Section 2.2) that the thickness of this sublayer is very important in determining whether or not the fluid flow produces erosion of the boundary surface. For the moment, let us simply note that the thickness of the sublayer, relative to the *boundary roughness* (Figure 2.4) has a marked influence on the equation for the velocity profile.

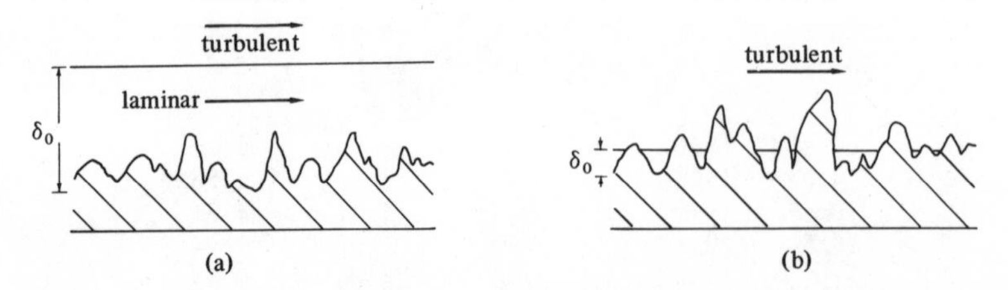

Figure 2.4. Definition of (a) smooth and (b) rough boundary surfaces.

Observations have shown that the velocity profile of air blowing over an *immobile* surface (see Figure A2.4 for a mobile surface) assumes the exponential form shown in Figure 2.5. The general equation of this curve may be expressed as

$$u = b(\ln z - \ln z_0) = b \ln\left(\frac{z}{z_0}\right) \tag{2.4}$$

where b is the slope of the curve, z is the distance from the boundary and z_0, as defined in Figure 2.5, is the thickness of a very thin layer of stationary fluid adjacent to the boundary. (Other forms of this equation are given in Appendix 2.3.) For hydrodynamically *rough* boundaries, Nikuradse has shown empirically (see Chow, 1959, p.202) that

$$z_0 = \tfrac{1}{30} d \,, \tag{2.5}$$

where d denotes the particle size for which 65% of the material is finer. Some experimental values of z_0 for various types of rough surface are given in Table 2.2. The possible role of a vegetation cover in reducing wind erosion is readily appreciated from these data. For *smooth*

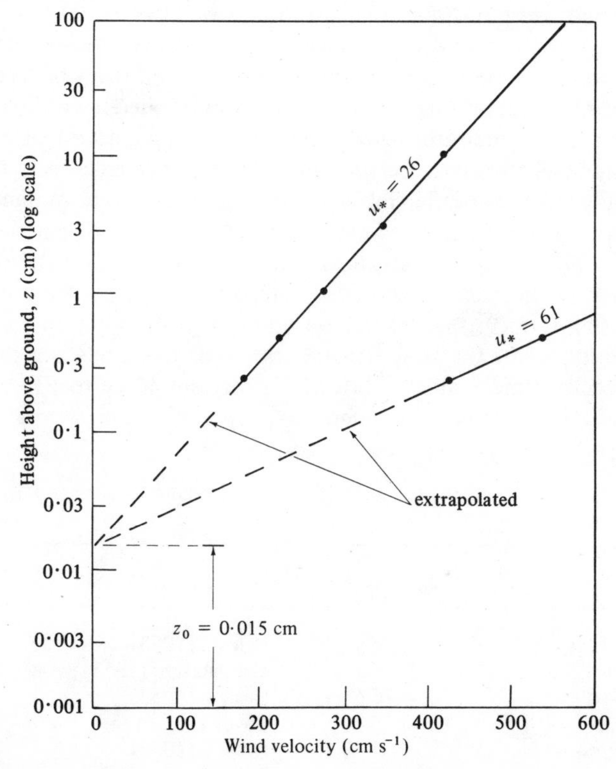

Figure 2.5. Velocity distribution in a wind tunnel (after Bagnold, 1953).

boundaries, it follows from Equation (2.3) that

$$z_0 \approx \frac{\nu}{9}\sqrt{\frac{\rho_f}{\tau_0}} \; . \tag{2.6}$$

So far, no rational interpretation has been offered for the coefficient b in Equation (2.4). The following treatment is based mostly on the ideas of the German physicist Ludwig Prandtl, one of the key figures in the early development of fluid mechanics.

Let us consider the factors controlling the shear stress between any two layers of flowing fluid. Newton long ago stated that, for *laminar* flow, the shear at any point in an ideal fluid (Chapter 3) is proportional to the momentum concentration ($\rho_f u$) gradient at that point, that is,

$$\tau = \nu \frac{d(\rho_f u)}{dz} = \nu \rho_f \frac{du}{dz} \; , \tag{2.7}$$

where the constant of proportionality is in fact the kinematic molecular viscosity. Shear is thus regarded as a flux of momentum from regions of higher velocity to regions of lower velocity. This is schematically shown in Figure 2.6, in which an imaginary interface is inserted within the fluid flow. If there were no molecular interference between the two sides of the interface, there would be no shear stress between the two layers of fluid. However, molecules are continuously moving from both directions across the interface, adding to the momentum of the slower layer and diminishing the momentum of the faster one. The amount of retardation of the faster layer, that is, the amount of shear resistance, is determined by the amount of molecular mixing ν and the contrast in momentum concentration (momentum per unit volume) between adjacent layers ($d\rho_f u/dz$). The fluid shear stress in laminar flow can, thus, be defined as the net flux of momentum across an interface within the flow, per unit area of that interface, that is, the net amount of momentum transferred across a unit area of the flow per unit time. It has dimensions of mass × velocity/time × area or $ML^{-1}T^{-2}$, corresponding to FL^{-2} in a system using force rather than mass.

Table 2.2. Roughness lengths for different types of boundary surface (from Sellers, 1965, p.150).

Type of surface	z_0 (cm)	Source
smooth mudflat	0·001	Deacon (1953)
dry lake bed	0·003	Vehrencamp (1951)
smooth desert	0·03	Deacon (1953)
5-6 cm grass	0·75	Covey *et al.* (1958)
60-70 cm grass	8-15	Deacon (1953)
500 cm fir forest	280	Baumgartner (1956)

The case of *turbulent* flow is only slightly different. The interaction between two layers of fluid in turbulent flow is no longer confined to molecular interference along the interface, but it also involves penetration of elements of fluid from one layer into the other. This eddy interference is also a form of viscosity and the symbol ϵ_m (epsilon) is termed the *kinematic eddy viscosity* coefficient. The subscript m denotes the mixing of momentum. As we shall see later, analogous mixing coefficients for other fluid properties (e.g. temperature, sediment concentration) exist, and these are not necessarily all equal. Unlike molecular viscosity, eddy viscosity is not a fixed property of a fluid, because it depends on the amount of turbulence; for this reason, no values of ϵ_m are given in Table 2.1. By analogy with Equation (2.7), the shear stress in turbulent flow is given by

$$\tau = (\nu + \epsilon_m)\frac{d(\rho_f u)}{dz} \approx \epsilon_m \frac{d(\rho_f u)}{dz} \approx \epsilon_m \rho_f \frac{du}{dz}, \qquad (2.8)$$

noting that ϵ_m is usually much bigger than ν. Prandtl, adopting an interpretation of fluid shear slightly different to that of Figure 2.6, reasoned that this eddy coefficient must be a function of the depth and frequency of penetration of fluid between layers, specifically that

$$\epsilon_m = l^2 \frac{du}{dz}, \qquad (2.9)$$

where l, the mixing-length, is a measure of the depth of eddy penetration, and the velocity gradient du/dz is linked to the frequency of penetration. Experimentally, it was shown by Schlichting, another German engineer, that, in the zone near to the boundary,

$$l = \kappa z, \qquad (2.10a)$$

where κ (kappa), the coefficient of proportionality, is the well-known von Kármán constant equal to 0·4, and z is the distance from the boundary

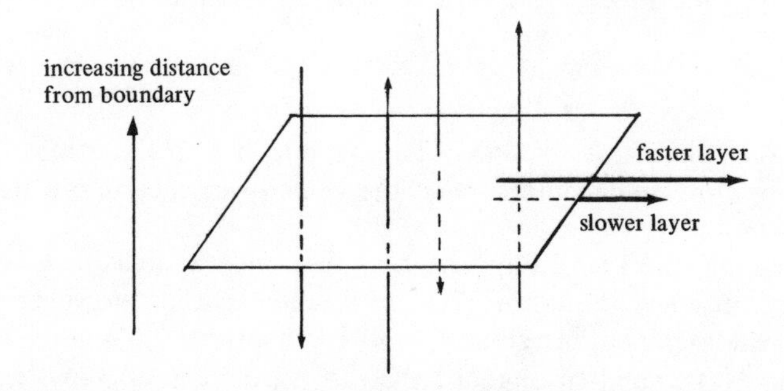

Figure 2.6. Schematic representation of kinematic viscosity in relation to the fluid shear stress.

surface. Combining Equations (2.8), (2.9), and (2.10a), we have

$$\tau = (\kappa z)^2 \rho_f \left(\frac{du}{dz}\right)^2 \tag{2.11}$$

or

$$\frac{du}{dz} = \frac{1}{\kappa}\frac{1}{z}\sqrt{\frac{\tau}{\rho_f}}$$

which on integration yields

$$u = \frac{1}{\kappa}\sqrt{\frac{\tau}{\rho_f}}\ln(z) + C\,. \tag{2.12}$$

Previously, we have noted that $u = 0$ and $\tau = \tau_0$ at $z = z_0$; substituting in Equation (2.12), we have

$$C = -\frac{1}{\kappa}\sqrt{\frac{\tau_0}{\rho_f}}\ln(z_0)\,.$$

If we restrict our attention to the flow *near the boundary surface*, we may assume that the fluid shear stress is equal to the boundary shear stress, that is,

$$\tau = \tau_0\,; \tag{2.13a}$$

substituting for C in Equation (2.12), we obtain

$$u = \frac{1}{\kappa}\sqrt{\frac{\tau_0}{\rho_f}}\ln\frac{z}{z_0}\,. \tag{2.14}$$

Comparison of this theoretical equation with the empirical velocity profile of Equation (2.4) indicates that the coefficient b (= $du/dz(\ln z)$] is given by

$$b = \frac{1}{\kappa}\sqrt{\frac{\tau_0}{\rho_f}}\,. \tag{2.15}$$

This should not be surprising. The dimensions of the term $\sqrt{\tau_0/\rho_f}$ are those of velocity, and indeed, the term is often represented in the form

$$\sqrt{\frac{\tau_0}{\rho_f}} = u_* \tag{2.16}$$

where u_* is called the *shear velocity*. The parameter b in Equation (2.4) is itself nothing more than a measure of the velocity of shearing in the fluid near the boundary.

Strictly speaking, the Prandtl approach applies only to areas near the fluid boundary, in which Equation (2.13a) is valid. Earlier, however, we observed that the exponential velocity profile [Equations (2.4) and (2.14)] appears to be valid throughout the *full* depth of fluid flow, not merely the boundary zone. The same point is noted by Rouse (1959, p.299) who attempts to reconcile this difference by suggesting that the mixing length is more appropriately given by

$$l = \kappa z \left(1 - \frac{z}{D}\right)^{1/2}, \tag{2.10b}$$

where D is the depth of flow, rather than by Equation (2.10a). [Naturally, as the boundary is approached and $z \rightarrow 0$, the two versions of Equation (2.10) merge.] The reasoning behind this suggestion by Rouse rests in the relation between the general fluid shear stress τ and the boundary shear stress τ_0. By definition, only at the boundary are these two identical [Equation (2.13a)], and the general relationship—used again in Appendix 2.10 and derived in Figure 3.20—is given by

$$\tau = \tau_0 \frac{D - z}{D}. \tag{2.13b}$$

Use of Equations (2.10b) and (2.13b) together, rather than (2.10a) and (2.13a), in the previous development of the velocity profile will again result in Equation (2.14), but this time applicable to the full depth of flow.

The reader may quite justifiably be rather frustrated by the speculative nature of many of these assumptions used in the development of the so-called logarithmic velocity distribution. In fact, however, as pointed out by Townsend (1956), the validity of this 'law of the wall' is only to a minor extent dependent upon the particular assumptions made about the mixing length, because most of the turbulent energy production takes place within a thin layer close to the wall in which the shear stress varies by less than 10%. Indeed, a combination of Equations (2.10a) and (2.13b) would also yield a theoretical velocity distribution quite close to that given by Equation (2.14), although rather more complicated. Hinze (1959, Chapter 7) provides a comprehensive and lucid discussion of these points.

The important point in this section is this: the velocity profile of a stream (or any fluid flow) provides a simple means of evaluating the boundary shear stress τ_0 acting within the fluid. This stress is, at equilibrium, equal to the drag exerted on the flow of fluid by the boundary per unit area of boundary surface. Now, just as the boundary exerts a drag on the flow of fluid past it, so the fluid flow exerts a drag, equal in magnitude but opposite in direction, on the boundary. This drag was named by DuBoys in 1879 as the *tractive force* of the fluid, and it has been widely used in attempting to explain fluid erosion on the beds of streams. The term 'unit tractive force' (denoted, just as boundary shear stress, by τ_0) is used here to describe the drag per unit area of the boundary.

2.2 Initiation of fluid erosion

The total tractive force acting along the bed and banks of a rectangular stream channel, over a length downstream of L, is given by (Figure 2.7)

$$T = \overline{\tau_0}(2D + W)L \tag{2.17}$$

where $\overline{\tau_0}$ is the *average* boundary shear stress over the channel boundary.

It is easily shown from Figure 2.7 that, for uniform flow,

$$\rho_f g D L W \sin i = \overline{\tau_0}(2D + W)L$$

and thus

$$\overline{\tau_0} = \gamma_w R s \tag{2.18}$$

where γ_w ($= \rho_f g$) is the unit weight of water, R is the hydraulic radius and s ($=\tan i = \sin i$ for small angles) is the water surface slope in radian measure. It must be emphasized that $\overline{\tau_0}$ is only an average value; it may vary considerably through the cross-section. The value of τ_0 over any particular unit area of the boundary can only really be evaluated in terms of the velocity gradient normal to that part of the boundary. Nevertheless, for wide channels, Equation (2.18) provides a reasonable estimate of the tractive force per unit area of the stream bed.

Experiments and theory both indicate that, for coarse bed material, the diameter of the largest particle that can be moved along the bed by fluid flow is a function of the unit tractive force on the bed. Expressed in a different way, the critical or threshold unit tractive force necessary to initiate bed load movement (τ_c) is a function of the size of the bed material. (As pointed out in Appendix 2.5, this is the basis for the much-quoted statement that the competence of a stream is a function of the sixth power of the channel velocity.) We shall begin by considering the approach adopted by White (1940). (Note that this approach is restricted to horizontal flow as indicated in Figure 2.8; for the more general case of flow over an inclined bed, as in Figure 2.14, the angle between the two force vectors in Figure 2.8 is smaller than 90° by an amount equal to the angle of slope.) Suppose that the channel bed is covered by n particles, of diameter d, per unit area. *If* all the bed shear is attributable entirely to these particles, each grain experiences a force of

$$F_D = \frac{\tau_0}{n} \tag{2.19}$$

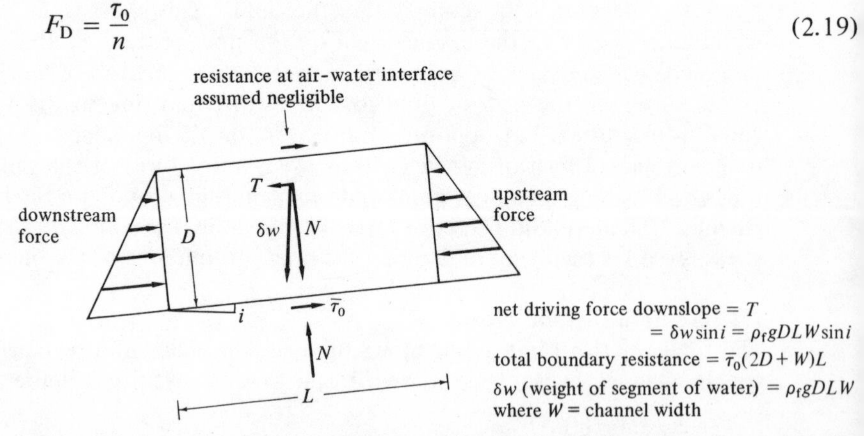

Figure 2.7. Free body diagram of segment of open channel flow through a rectangular channel.

in the direction of fluid flow; for rough boundary conditions this force passes through the centre of gravity of the particle. If we consider the moments acting on any grain about its downstream contact point (Figure 2.8), at limiting equilibrium, the moment due to the (submerged) weight of the grain is just balanced by the moment due to the drag force; this yields

$$\tau_c = \rho' g d \eta (\pi/6) \tan\phi \tag{2.20}$$

where $\rho' g$ (= $\rho_s g - \rho_f g$) is the submerged unit weight of the individual particles (ρ_s being the actual mass density of a particle) and η (eta) (= nd^2) is a dimensionless measure of the packing of the particles on the bed [$\eta(\pi/4)$ being the fraction of the bed area covered by spherical grains]. White's experiments revealed that, for rough surfaces (particles projecting up through the laminar sublayer), measured values of τ_c were, in fact, about half of those predicted by Equation (2.20). He attributed this to the observation that, in turbulent flow, pulsations in velocity produce *maximum* drag values equal to about twice the *mean* (over time) value.

Equation (2.20) has been expressed in many ways by workers both before and after White. Shields (1936), for instance, termed the ratio $\tau_c/\rho' g d$ the entrainment function and, in a now classic diagram (Figure 2.9), plotted experimental values of this function against the ratio d/δ_0. Note that for fully turbulent flow ($d/\delta_0 \rightarrow \infty$), the entrainment function assumes a constant value of about 0·06. Under conditions of laminar flow at the bed, the entrainment function is shown to increase as $d/\delta_0 \rightarrow 0$.

Bagnold (1953) provides data for the onset of wind erosion (in wind-tunnels) over a flat surface which are similar to Shields' data for the entrainment function in water. Figure 2.10 shows a plot of critical shear

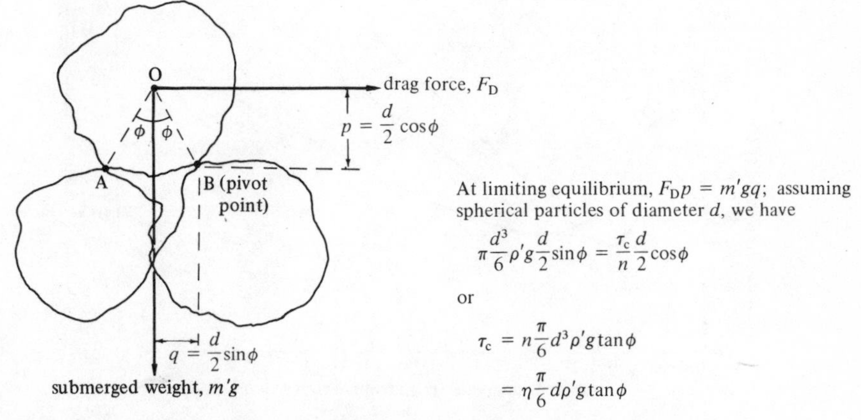

Figure 2.8. Moments on a submerged sand particle resting on a horizontal flat bed at the threshold of movement, according to White (1940).

velocity against the square root of particle size on the boundary. For particles larger than about 1 mm in diameter, the relationship may be expressed in the form

$$u_{*_c} = bd^{1/2} \tag{2.21a}$$

where b is the regression coefficient and is, of course, constant. Alternatively, we may define the relationship in the form

$$u_{*_c} = A\left(\frac{\rho'}{\rho_f}g\right)^{1/2} d^{1/2} \tag{2.21b}$$

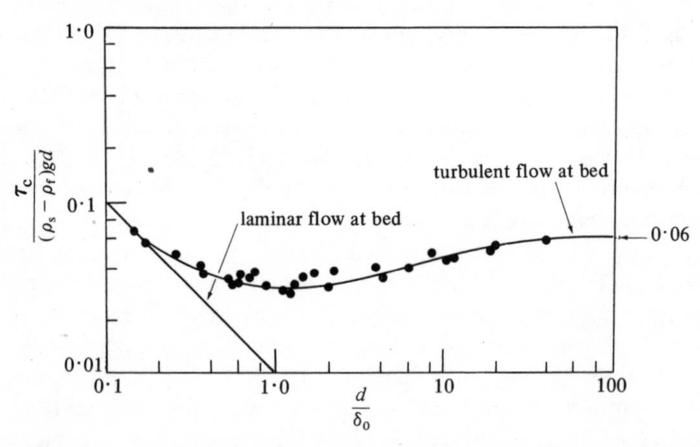

Figure 2.9. Shields' diagram relating the entrainment function to bed roughness under conditions of a plane bed surface (after Shields, 1936).

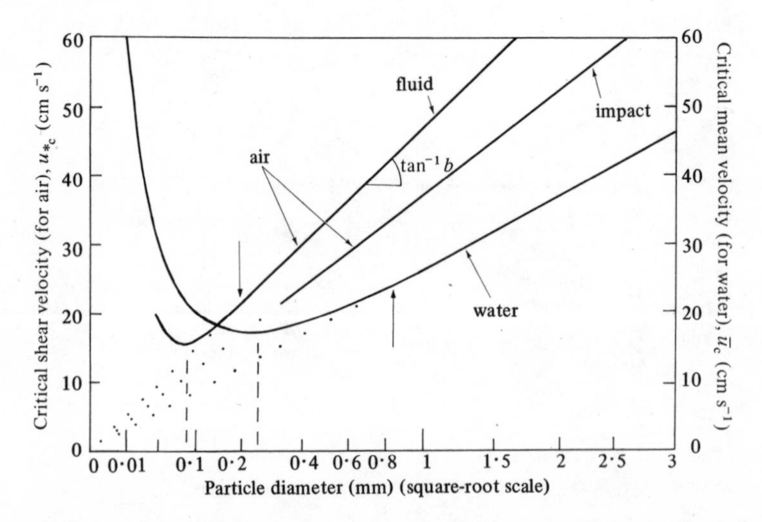

Figure 2.10. Variation of critical velocity with particle size for water and air (after Hjulström, 1935 and Bagnold, 1953).

in which A is a similar, but dimensionless, coefficient; assuming ρ', ρ_f, and g are constant, A is also constant for particles coarser than 1 mm. From the data it is found that A is equal to 0·1. If we square both sides of Equation (2.21b) and substitute for u_{*_c} in terms of τ_c we obtain

$$\frac{\tau_c}{\rho' g d} = A^2 \qquad (2.22)$$

in which the left-hand side is Shields' entrainment function.

By comparison with Equation (2.20), it also follows that Bagnold's A^2 should be equal to $\eta(\pi/6)\tan\phi$ for spherical grains, provided that boundary conditions are rough. If these conditions are not met, a significant part of the fluid force on individual particles will then be skin friction drag, rather than form drag, and the force vector will pass through the particle at a small height above the centre of mass. As pointed out by Leliavsky (1966, p.57), this factor may easily be incorporated into the solution by adding a dimensionless coefficient α into the right-hand side of Equations (2.20) and (2.21), with $\alpha \rightarrow 1$ as boundary conditions become rough.

Comparing Bagnold's data with Shields' diagram, it appears that A^2 is higher for water (0·04–0·06) than for air (0·01) under rough conditions. This represents something of a puzzle because A appears to depend solely on η, ϕ, and particle shape ($\pi/6$). It is possible, however, that this merely represents unnoticed differences in test conditions. Equations (2.20) and (2.21) are restricted to flat bed surfaces. It has been observed that Shields' entrainment function is strongly influenced by bed form (for a rippled surface values of 0·1 to 0·25 are typical) and the small difference between the data of Bagnold and Shields for a flat bed may be merely experimental error.

The reader should be warned that the theoretical approach of White, as embodied in Figure 2.8, suffers from a further problem in addition to those already noted (this applies also to the empirical data of Bagnold and Shields). Figure 2.8, and thus Equation (2.20), describes an *average* condition. The angle AOB, defined by 2ϕ, may vary from particle to particle. Moreover, even if ϕ were constant, there is no reason why AB should be parallel to the drag force vector for every particle. Again, this is an average condition; individual particles will deviate from this state, depending on the randomness of the packing on the fluid boundary. Accordingly, Equation (2.20) cannot predict the critical unit tractive force necessary to initiate the movement of the *first* particle; it relates to the stress needed to move a hypothetical *average* particle. Similarly, the experiments reported by Bagnold and Shields suffer from the practical difficulty of defining the actual threshold of debris movement; this may account for some of the discrepancies in the data.

Both Figures 2.9 and 2.10 illustrate a very important point in relation to the initiation of fluid erosion. For particles smaller than a critical size (for d/δ_0 in 2.10) the value of A (or the value of the entrainment function) begins to increase, rather than decrease, as particles get smaller. Equations (2.20) and (2.21) (with A constant) are thus valid only for particles larger than 0·2 mm for air and about 0·8 mm for water; for smaller particles, a larger unit tractive force (relative to particle weight) than that predicted by this equation is needed. Indeed for very fine debris ($d < 0\cdot05$ mm for air and $d < 0\cdot3$ mm for water) there is an *inverse* relationship between particle size and critical unit tractive force; as the particle gets smaller, a larger unit force is required to initiate its movement. A little thought will reveal that this is not entirely surprising. Equation (2.20) is based on the assumption that the fluid flow around the bed particles is turbulent. Now, we have already noted that, adjacent to the boundary of fluid flow, there is a thin sublayer in which laminar conditions prevail; further, from Equation (2.3), we know that the thickness of this sublayer is inversely related to the boundary shear stress. Equations (2.3) and (2.20) are combined graphically in Figure 2.11a. Two changes are shown to occur as the boundary shear stress is decreased: (a) the largest size of particle that can be moved is reduced, and (b) the thickness of the laminar sublayer increases. There is thus a critical size of particle for which, at the moment when the *predicted* unit tractive force is just great enough to initiate movement, the particle no longer projects

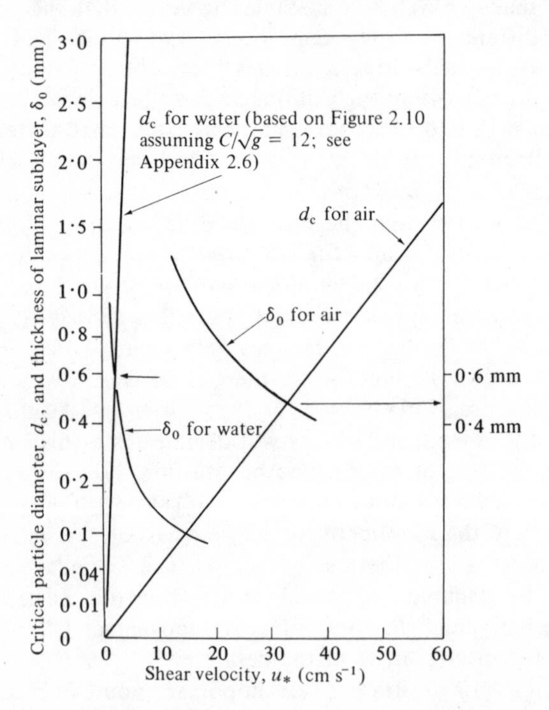

Figure 2.11a. Critical particle size and thickness of laminar sublayer as a function of the shear velocity.

into the turbulent flow. In Appendix 2.6, it is shown that this critical size is given by

$$d^3 = \frac{\rho_f}{\rho'} \frac{(11 \cdot 6)^2 \nu^2}{g A^2} \tag{2.23}$$

which reduces to $d = \delta_0 \approx 0 \cdot 5$ mm and $d = \delta_0 \approx 0 \cdot 7$ mm, for air and water respectively. These values agree reasonably well with the experimental values for the lower limit of the square root relationship between d and u_{*_c} (Figure 2.10) obtained by Bagnold and assembled by Hjulström (1935). For particles smaller than this critical size (Figure 2.11b) it is easy to appreciate that a larger unit tractive force is necessary to initiate movement because the particle is submerged in the laminar sublayer. (The exact mechanics of particle movement in this sublayer are still uncertain.) This is borne out in Shields' data also: as particles become more submerged in the sublayer ($d/\delta_0 < 1$), the entrainment function increases. Note, finally, that the fact that entrainment does occur for d/δ_0 values less than unity indicates that it is not necessary to increase the unit tractive force until the sublayer is thinned below d, in order to begin movement.

One serious criticism of the tractive force approach to movement of debris along, or from, the boundary of fluid flow is that it neglects the force acting on particles *normal* to the main flow. Both theory and experimental evidence indicate the presence of a *lift force* acting away from the boundary (upwards on flat ground) of fluid flow. Jeffreys (1929), for instance, deduced this working from the Bernoulli equation (Appendix 2.7), and a concise summary of his ideas is provided by Leliavsky (1966, p.66). The argument is as follows. Fluid flow past a particle on a level boundary is faster over the top than around the sides, and, indeed, at the contact between the particle and boundary, the fluid is stagnant. From the Bernoulli principle it follows that the fluid pressure is greater on the underneath side of the particle than on the upper side

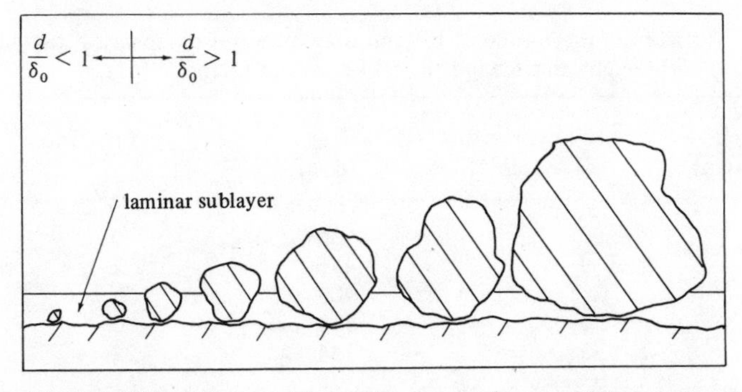

Figure 2.11b. Influence of particle size on flow conditions around a particle on fluid boundary.

(Figure 2.12) and, as a consequence, a net upward force is exerted on the particle. Early experiments by White (1940) suggested that this pressure difference is very small, but more recent work, using more refined techniques, has shown this conclusion to be wrong. Chepil (1961), using a technique illustrated in Figure 2.12b, demonstrated that at levels very near to the boundary (Table 2.3) the lift force on particles (d = 0·8 cm) is comparable in magnitude to the drag force. As particles are moved away from the bed, the lift force decreases very rapidly and, for these

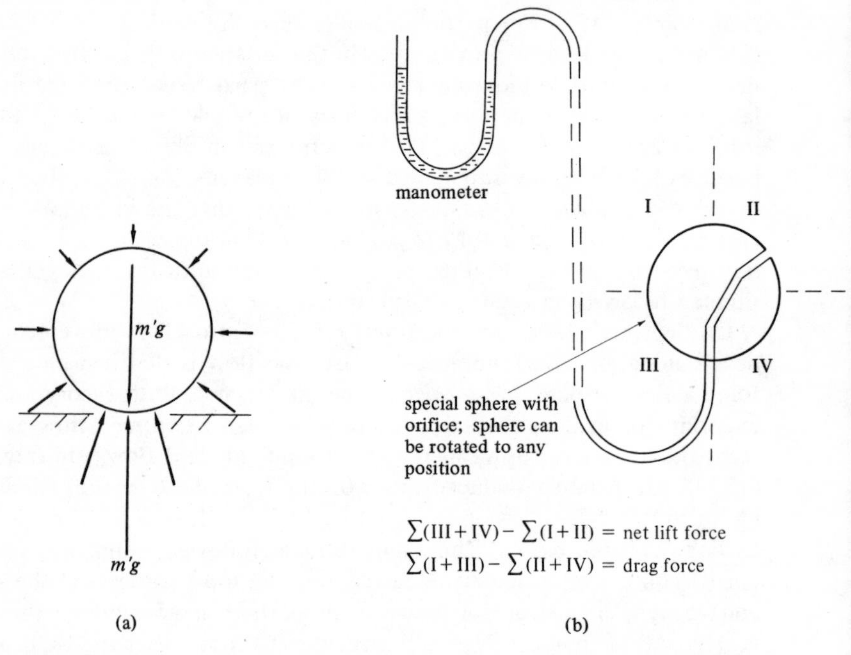

Figure 2.12. (a) Pressure (excluding drag) on a grain submerged in fluid flow; (b) Chepil's method of measuring total pressure at any point on a submerged sphere.

Table 2.3. Variation in magnitude of lift and drag pressures on spherical particles (d = 0·8 cm) with height above wind tunnel bed (from Chepil, 1961).

Height of sphere above bed (cm)	Net lift pressure (dynes cm^{-2})	Drag (dynes cm^{-2})	Lift/Drag
2·5	0	530	0
2·0	9	476	2
1·5	21	408	5
1·0	20	318	6
0·5	33	144	23
0·25	33	78	41
0·0	34	35	97

particular data, the lift force had disappeared completely at a distance of 2·5 cm from the bed. This rapid fall-off in the strength of the lift force is presumably related to the rapid decrease in the velocity gradient (Figure 2.2), and therefore in the pressure gradient, at increasing distances from the boundary surface. Whatever the full explanation for the phenomenon, however, it has very important implications for sediment transport, as we shall note shortly.

The importance of the data shown in Table 2.3, and similar results by other workers (e.g. Einstein and El-Samni, 1949), is that they demonstrate that, *irrespective of the magnitude of the tractive force*, the lift force is capable of initiating transport of loose debris on the boundary of fluid flow. As soon as the lift force becomes larger than the submerged weight of a particle, the particle will move up from the bed and become incorporated into the flow of the fluid. Whether or not the lift force is the key mechanism in initiating sediment transport along the boundary surface is difficult to judge, because the lift force and the tractive force on a particle are very closely interrelated. We have already noted [Equation (2.19)] that the drag force on an individual particle is linked to the average boundary shear stress and the size of the particle:

$$F_D = \frac{\tau_0}{n} = \frac{\tau_0 d^2}{\eta},$$

where η is a measure of the intensity of packing. Furthermore, from Equation (2.16), we know that the boundary shear stress is given by

$$\tau_0 = \rho_f u_*^2$$

where u_* is the shear velocity. The term u_* is, itself, a function of the velocity of the fluid and, thus, combining all these points, we obtain an expression of the form:

$$F_D = Ca\frac{\rho_f}{2}u_0^2, \tag{2.24}$$

where a is the surface area of the particle, u_0 is the velocity of the fluid surrounding the particle and C is a proportionality coefficient. For isolated particles in a moving fluid, the coefficient C is a function of the shape of the particle (see Appendix 2.8); it is termed the *drag* coefficient and written C_D. For close packing of bed material there is interference among the particles and C is then more a measure of packing (as noted by White) than the shape of an individual particle. Experiments (e.g. Einstein and El-Samni, 1949) have shown that an analogous expression may be used for the lift force:

$$F_L = C_L a\frac{\rho_f}{2}u_0^2, \tag{2.25}$$

where C_L is defined as the *lift coefficient.* Thus, both lift force and drag force increase with the square of the velocity and, in order to establish the relative importance of the two forces in initiating particle movement, very precise measurements are necessary.

The lift force approach certainly provides a very logical explanation for the movement of particles by *saltation* (Figure 2.13a) along a fluid boundary. If, temporarily, we neglect the drag force on a particle, it is easy to show that, at high velocities, particles would be expected to move up and down in a small thickness of fluid adjacent to the bed. The reason is, as demonstrated in Table 2.3, that the lift force rapidly weakens with increasing height above the bed. *At* the bed, the lift force, at high velocities, may be greater than the submerged weight of the particle and, in accordance with Newton's second law of motion, the particle will accelerate upwards. As the particle moves away from the bed, the lift force decreases and, when lift is smaller than the submerged weight, deceleration, and eventual acceleration back towards the bed, begin. Thus, in the absence of a drag force, a particle would be expected to oscillate continually about a critical height at which the lift force and the submerged weight of the particle are equal (Figure 2.13b). If we now superimpose onto this motion the drag force acting on the particle, the result is a bouncing motion in the direction of fluid flow, as shown in Figure 2.13a. High-speed photographs show that the initial direction of movement in the characteristic path of a saltating particle is, in fact, normal to the bed, thus adding support to the idea that the lift force is the actual mechanism initiating particle movement.

Actually there is an important difference between saltation of sand particles in air and in water. As soon as saltation begins in air, *subsequent* particle movement is attributable, not so much to the lift force on a static particle, as to the *impact* of other particles hitting the bed. This has been very neatly demonstrated in wind-tunnel experiments by Bagnold (1953). As already noted, in order to initiate saltation of bed material, the bed velocity must be built up to a certain critical value dependent on the

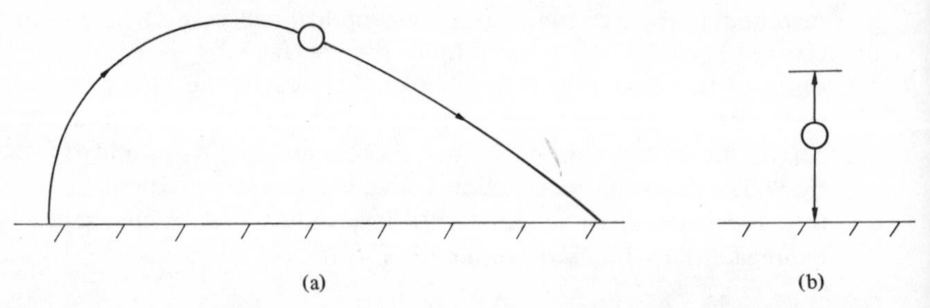

Figure 2.13. Saltation: (a) characteristic path of a saltating particle along the fluid bed; (b) hypothetical movement of a sand particle due to lift force (in absence of drag force).

particle size of the material. Bagnold defines this value as the *fluid threshold.* As soon as saltation begins, there is a small drop in the bed velocity because of the extra drag on the flow caused by continually picking up material from the bed. The surprising point is that saltation (in air) *continues* to take place even though the bed velocity falls below the fluid threshold value. The reason for this is that, despite the fall in the lift force resulting from the fall in bed velocity, the impact of grains already in motion enables saltation to continue. The bed velocity could, in fact, be lowered even further and saltation would still continue, although naturally there is a limit to this. Bagnold defines this limit as the *impact threshold* (Figure 2.10); for coarse material, the impact threshold also follows Equation (2.21), although the value of A is lower, being 0·085 compared to 0·1 for the fluid threshold.

The question remains: Why does this happen in air but not in water? The answer is attributable to the fact that, whereas the submerged density of sand particles is about a thousand times the density of air at atmospheric pressure, it is less than twice the density of water. The important point is this. The impact of a moving sand particle with a stationary grain on the boundary surface is essentially the same as the impact of numerous particles of fluid on that grain (as pointed out in Appendix 2.8, this is, in fact, the meaning of form drag); but, because the density of a sand particle is so much greater than the density of an equivalent air particle, the equivalent impact force of a saltating sand particle is much greater than that of an air particle. Hence, the solid-to-solid impact force is sufficient to maintain saltation at velocities where the air-to-solid impact force is too small; expressed in another way, a much smaller bed velocity is needed for the threshold condition with solid-to-solid impact than with air-to-solid impact. On the other hand, the difference between the submerged density of a sand particle and the density of water is so small that the difference in threshold bed velocity between solid-to-solid impact and water-to-solid impact is negligible.

Saltation is not the only process by which fluids erode debris from the fluid boundary and, indeed, saltation is restricted primarily to particles of sand size. Slightly coarser debris may be moved by *surface creep* in wind, and much coarser debris may be moved by water via *rolling* along the bed. The term surface creep is used by Bagnold when the impact of a moving sand particle is insufficient to propel a stationary particle into the air but nudges it forward along the bed. Experiments have shown that this mechanism commonly accounts for as much as a quarter of total bed-load movement during wind erosion. Because the impact force is so small, relatively, in water, surface creep is not an important process here. On the other hand, rolling of particles may be very important in fluvial (water) erosion, but is a minor mechanism in wind erosion. Rolling is the primary means by which very coarse debris (from gravel to boulders) is moved along the bed of a stream channel. Although it is impossible to develop lift forces large enough to produce saltation of such heavy material, sufficient tractive force is often available to produce rolling. This is particularly the case (Figure 2.14) in steeply-sloping channels. Here the drag force is supplemented by the downslope component of the particle's weight; moreover, the frictional resistance between particle and boundary (proportional to the component of particle weight normal to the flow) is also reduced. Wind erosion of coarse material by rolling is rare because, even at high velocities, the small mass density of air prevents very high tractive forces from developing [Equation (2.24)]. *Suspension*, which occurs in both air and water, may be regarded as a simple extension of saltation in which the ratio of lift force to submerged weight on particles is so great that particles remain suspended above the fluid boundary for as long as the flow conditions prevail. The lift forces involved in suspension are, however, rather different to those described for saltation and are dependent upon turbulent flow conditions, whereas the lift force approach to saltation is also valid for laminar flow. Data provided by White (1940) indicate that the threshold shear velocity necessary to initiate suspension of sediment is given by

$$u_{*_c} = \frac{1}{7}\omega \tag{2.21c}$$

where ω is the terminal fall velocity of the sediment particles. The mechanics of suspension are fully discussed in the next section.

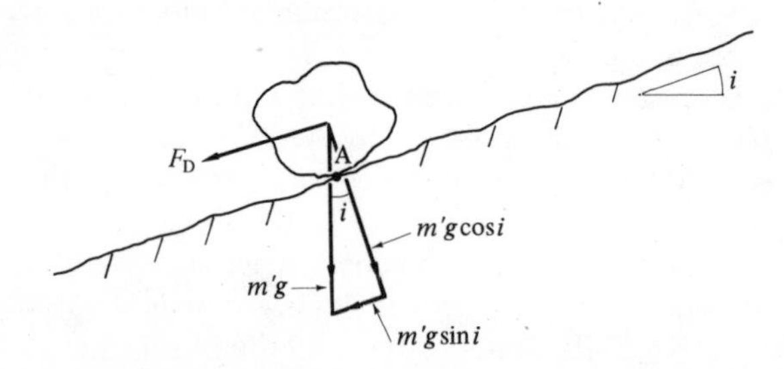

Total downslope force $= F_D + m'g\sin i$

Total frictional resistance at A $= \mu m'g\cos i$

Figure 2.14. Forces on a submerged particle on a sloping bed; m' is submerged mass of boulder.

2.3 Rates of fluid erosion

Just as there is a limit to the *size* of particles that can be moved by fluid flow (competence) according to the energy of the flow, similarly there is a limit (capacity) to the *amount* of material that can be moved per unit time. In this section we shall examine the relationship between capacity and flow conditions for both bed load and suspended load. First, however, let us emphasize that it is only under infrequent conditions that the maximum transport capacity of fluid flow is ever attained. Most processes of fluid erosion in nature are limited by the rate at which weathering is able to make loose debris available for transport, rather than by the energy of the transporting medium. This is strikingly revealed by the data in Table 2.4 showing the fall-off in sediment produced by hillslope erosion in a simulated rainstorm. Secondly, it should be noted that the term capacity is only meaningful in relation to a specified size-class of debris. In natural fluid flow, involving a wide range of particle size, it is quite possible to have the transporting medium at capacity for one size of debris and under capacity for another size. Most formulae for the capacity of sediment transport by fluids are based on the simplifying assumption, or laboratory condition, of uniform debris size.

A neat treatment of the factors controlling maximum wind erosion rates has been provided by Bagnold (1953, pp.64-74), and we shall begin our discussion of bed load capacity using this approach. Consider a single sand grain lifted from the ground surface with *horizontal* velocity u_1, travelling a distance L horizontally, and then striking the surface again with a *horizontal* velocity u_2, all of which is lost on impact. This grain must absorb momentum from the air. The actual amount absorbed is equal to $m(u_2 - u_1)$, where m is the mass of the particle; the amount absorbed per unit length of ground travelled is $m(u_2 - u_1)/L$. Now consider a collection of particles moving in saltation along a track of unit width and passing across a fixed line in unit time. The rate of loss of momentum per unit length per unit time is given by

$$p = \frac{Nm}{t}\frac{u_2 - u_1}{L}, \tag{2.26}$$

where N is the number of particles passing the line in time t. If q_s is defined as the collective mass of saltating particles passing a unit width of the line per unit time, Equation (2.26) may be written

$$p = q_s \frac{u_2 - u_1}{L}. \tag{2.27}$$

Table 2.4. Decrease in sediment transport rate during erosion of hillslope plot under artificial rainfall supply (from Emmett, 1970, p.A42).

Date (1967)	Time from start of rainfall application (min)	Sediment concentration[a] (mg litre^{-1})
29 July	24	288
29 July	35	184
29 July	82	41
29 July	119	36
30 July	15	160
31 July	150	49

[a] total sediment: inorganic and inorganic

Note that the dimensions of p are force/area (M L^{-1} T^{-2}). In fact, p is the resistance (as a stress) exerted on the air flow by every unit area of the boundary due to sand particles being picked up. Thus p represents *part* of the boundary shear stress. The other part is resistance provided by stationary particles on the boundary. Because the mass density of a sand particle is so much greater than *air*, Bagnold assumes that the loss of momentum due to saltating sand particles accounts for virtually all the boundary shear stress during *wind* erosion, so that $p = \tau_0$. Further, because the initial path of a saltating particle is almost vertical, u_1 can be taken as zero. Accordingly Equation (2.27) reduces to

$$q_s = \frac{L}{u_2}\tau_0 \,. \tag{2.28}$$

The ratio L/u_2 is simply a description of certain aspects of the characteristic path of a saltating grain; experiments have shown that it is, in fact, equal to v_1/g, where v_1 is the initial vertical velocity of the saltating grain. Bagnold now makes the plausible assumption that the initial vertical velocity is proportional to the shear velocity according to the equation

$$v_1 = Bu_* \,, \tag{2.29}$$

where B is defined as the impact coefficient. Equation (2.28) then becomes

$$q_s = \frac{Bu_*}{g}\tau_0 \,;$$

using $\tau_0 = \rho_f u_*^2$, we obtain

$$q_s = B\frac{\rho_f}{g}u_*^3 \,. \tag{2.30}$$

The total bed load movement in air q_b is composed of surface creep q_c as well as saltation q_s. Experiments by both Bagnold (1953) and Chepil (1945) have shown that q_c is approximately $\frac{1}{3}q_s$; on the assumption that this is generally valid, the total bed load movement is given by

$$q_b = \tfrac{4}{3}B\frac{\rho_f}{g}u_*^3 \,, \tag{2.31}$$

which can also be written (Appendix 2.9) in the form

$$q_b = \alpha \times \tfrac{4}{3}B\frac{\rho_f}{g}(u-u_t)^3 \tag{2.32}$$

where u_t is the threshold velocity at a height k' above the bed, u is the actual velocity at height z, and α is a constant. In words, the bed load discharge, at capacity, is a function of the cube of the excess of the wind velocity over the threshold velocity. We shall now see that, notwithstanding differences in transport mechanisms, this deductive-empirical formula for air is similar to many for bed load transport in stream channels.

Numerous formulae for bed-load movement in stream channels exist. One of the oldest is that deduced by DuBoys:

$$q_b = c\tau_0(\tau_0 - \tau_c) \tag{2.33}$$

where c is a variable coefficient. In its emphasis on the excess of the boundary shear stress over the critical stress, it is not unlike Bagnold's expression for the case of air. Similar formulae have been provided by various workers and the reader is referred to Leliavsky (1966) for a comprehensive discussion. The best known bed load formulae are those of Hans Einstein. In his 1942 paper, he developed a formula for uniform bed material; and in 1950 he extended it to material of mixed sizes. Although the later formula is much more complicated, it is based on the same approach as that used in the 1942 paper, which, for simplicity, we shall discuss here.

Consider a stretch of channel bed of unit width along which sand particles are saltating with an average horizontal leap of L. In Figure 2.15, all particles dislodged within a length L upstream from AB will pass over this line; none originating from further upstream will do so. If the bed load rate is defined as the *number* of particles passing AB per second, this is equal to the number of grains dislodged over the length L per second. This is given by

$$\frac{N}{t} = \frac{L}{A_1 d^2} p_s$$

where $A_1 d^2$ is the bed area occupied by each grain, $L/A_1 d^2$ is the number of grains on the bed over a length L and unit width, and p_s is the probability of a grain being dislodged in a second. If the volume of each grain is represented by $A_2 d^3$, the *total mass* of grains crossing AB every second is given by

$$q_b = Lp_s \frac{A_2 d^3}{A_1 d^2} \rho_s$$

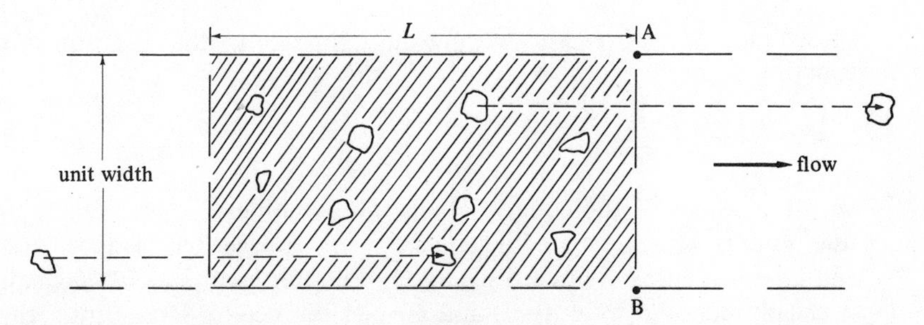

Figure 2.15. Definition diagram (plan) for Einstein's analysis of bed load movement.

or

$$q_{\mathrm{b}} = \frac{A_2}{A_1} L d \rho_{\mathrm{s}} p_{\mathrm{s}} \,. \tag{2.34}$$

The problem is to evaluate p_{s}. Einstein (1942) postulated, quite reasonably, that this probability is a function of the ratio of the lift force F_{L}, to the submerged weight of a particle w'; using [from Equation (2.25)]

$$F_{\mathrm{L}} = C_{\mathrm{L}} \frac{\rho_{\mathrm{f}}}{2} u_0^2 A_1 d^2 = \mathrm{f}(d^2 \tau_0) \,,$$

and

$$w' = (\rho_{\mathrm{s}} - \rho_{\mathrm{f}}) g A_2 d^3 = \mathrm{f}[(\rho_{\mathrm{s}} - \rho_{\mathrm{f}}) g d^3] \,,$$

we obtain

$$p_{\mathrm{s}} = \mathrm{f}\left[\frac{\tau_0 d^2}{(\rho_{\mathrm{s}} - \rho_{\mathrm{f}}) g d^3}\right]$$

or

$$p_{\mathrm{s}} = \mathrm{f}\left(\frac{\tau_0}{\rho' g d}\right) . \tag{2.35}$$

The right-hand side of Equation (2.36) is dimensionless, whereas p_{s} has dimensions of the reciprocal of time (T^{-1}). The right-hand side should, therefore, be divided by an appropriate time interval. Einstein argues that the appropriate interval is the time taken for a grain to fall a distance equal to its own diameter d at its *fall velocity* ω (omega). Equation (2.35) then becomes

$$p_{\mathrm{s}} \frac{d}{\omega} = \mathrm{f}\left(\frac{\tau_0}{\rho' g d}\right) ;$$

combining this with Equation (2.35), we obtain

$$\frac{q_{\mathrm{b}} d}{L d \rho_{\mathrm{s}} \omega} = \mathrm{f}\left(\frac{\tau_0}{\rho' g d}\right)$$

or, assuming $L = \mathrm{f}(d)$, and writing the right side of the equation in power function form,

$$\frac{(q_{\mathrm{b}}/\rho_{\mathrm{s}})}{d\omega} = a\left(\frac{\tau_0}{\rho' g d}\right)^b , \tag{2.36a}$$

with both sides of the equation being dimensionless. (The term $q_{\mathrm{b}}/\rho_{\mathrm{s}}$ used here is equivalent to Einstein's q_{b}; q_{b} is used here as mass discharge and by Einstein as volume discharge.) The left-hand side of this equation is usually denoted by Φ (phi) and termed the bed-load function; the right-hand side, denoted by $1/\Psi$ (psi), approaches Shields' entrainment function towards the moment of initiation of bed-load movement. Of course, this approach is not entirely deductive; empirical evidence is needed to ascertain the appropriate constants in the relationship $\Phi = a(1/\Psi)^b$. So far, however, no unique set of values has emerged, although it is commonly accepted (following Brown, 1950) that $b = 3$. In this case, the bed load equation becomes

$$\frac{q_b}{\rho_s} = a\omega d\left(\frac{\tau_0}{\rho' g d}\right)^3 \tag{2.36b}$$

or, on substituting $\omega \propto d^{1/2}$ (which is valid for particles coarser than $d = 0{\cdot}062$ mm, and thus appropriate for most bed material), and expressing the tractive stress in terms of the depth-slope product,

$$q_b \propto \frac{(Rs)^3}{d^{1/2}} \; . \tag{2.36c}$$

Other bed load formulae are discussed by Henderson (1966). Almost all of them are expressed in the form of functional relationships in which the values of the constants are determined empirically from flume or wind-tunnel experiments. As a consequence they are really restricted to debris which is small enough to be accommodated in laboratory models; whether or not the same relationships hold for coarser material in natural channels is uncertain. Unfortunately, the immense problem of measuring bed load discharge in natural channels renders it very difficult to test these flume-based relationships accurately.

Turning to material transported in suspension, either in air or water, there is again no satisfactory equation for predicting sediment discharge in terms of flow conditions. Several equations exist for predicting the gradient in sediment concentration with height above the boundary surface, but in order to yield an absolute value of sediment load, integrated over height, most of these equations demand field measurements of sediment concentration to be made at some height above the boundary surface. Einstein's (1950) formula is an exception to this statement. The suspended load of wind or water is so highly influenced in nature by the availability of sediment, however, that it is perhaps meaningless to attempt to predict sediment discharge purely in terms of flow conditions. It is well established that the severity of wind erosion is strongly affected by the dampness of the ground surface. The same remark is true for erosion by overland flow on hillslopes. The two major sources for suspended sediment in stream channels are usually river-bank collapse and valley-side erosion; neither supplies sediment at a rate determined by the flow conditions in the main channel network although, obviously, there will be some correlation. Moreover, much of the suspended material in stream channels is so fine that almost any condition of flow will transport it provided that some mechanism exists for getting this material into suspension. The rate of suspended sediment discharge is, therefore, probably best analyzed in terms of sediment supply rather than stream capacity, and, in turn, this is a function of many processes, some of which are dealt with in other chapters.

In the remainder of this section, we shall examine the factors controlling the relative concentration of suspended sediment at different heights above the fluid boundary. Consider a small area of fluid in a plane parallel to the boundary (Figure 2.16). In turbulent flow, some currents of fluid are moving up through this plane and some are moving down; the plane itself may be visualized as moving in the direction of flow of the fluid. At equilibrium, the same amount of fluid must move up through the plane as down in unit time. Both sets of currents are transporting sediment through the plane. If this sediment had no weight, complete mixing would take place and a uniform concentration of sediment would result. However, because these particles do have weight, they are continuously settling (at a terminal fall velocity ω) *within* the moving fluid, and this must modify the gradient of sediment concentration. The effect is to concentrate more sediment nearer the bed. Our task is to determine, quantitatively, this concentration gradient.

Because the concentration of sediment is greater beneath the plane than above it, more sediment must be transported up through the plane than is transported down, if, as assumed above, the same volume of fluid moves up as down. In attempting to determine the magnitude of this upward flux of sediment, it is useful to recall the previous discussion (Section 2.1) on the nature of the internal fluid shear stress.

In Figure 2.6, fluid shear was depicted as an exchange of momentum between adjacent layers of the flow due to transverse eddy movements. The fluid shear stress is, in fact, commonly defined as the net *flux* of momentum *through* a layer of fluid of unit area, towards the boundary, per unit time, as we have already noted. The magnitude of this flux is given by the gradient in momentum concentration $\mathrm{d}\rho_f u/\mathrm{d}z$, and a mixing coefficient ϵ_m. By analogy with Equation (2.8) it is easy to visualize the net flux of sediment upwards (away from the boundary), through a unit area of fluid per unit time, as an expression of the form

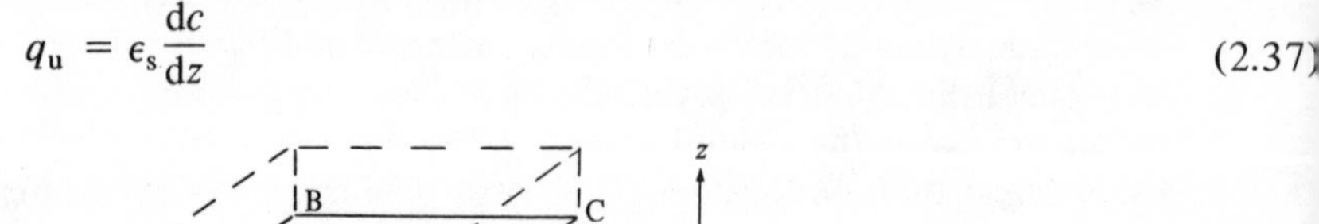

$$q_u = \epsilon_s \frac{\mathrm{d}c}{\mathrm{d}z} \qquad (2.37)$$

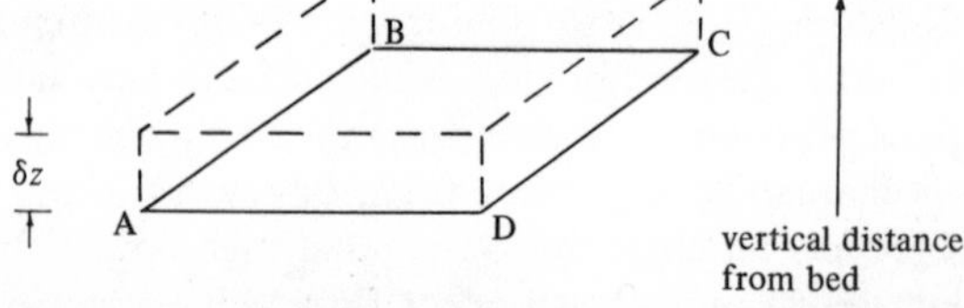

Figure 2.16. Definition diagram for von Kármán-Prandtl approach to suspended sediment distribution in open channel flow; δz is the thickness of a layer in which, in unit time, sediment with fall velocity ω moves past ABCD (unit area).

where $\mathrm{d}c/\mathrm{d}z$ is the gradient in sediment concentration and ϵ_s is comparable to, but not necessarily equal to, ϵ_m. This upward flux is balanced by the settling of particles under their own weight; if we consider the fluid above the plane ABCD, with particles at a concentration c settling at a terminal fall velocity ω, the rate of downward movement is given by

$$q_d = -c\omega \,. \tag{2.38}$$

At equilibrium, therefore, we have $q_u - q_d = 0$, or

$$\epsilon_s \frac{\mathrm{d}c}{\mathrm{d}z} = -c\omega \tag{2.39}$$

which, on integration, yields the concentration c as a function of the height above the bed z.

The solution of Equation (2.39) varies according to the assumptions made regarding the mixing coefficient ϵ_s. The simplest case is when ϵ_s is taken as constant (independent of z); this yields (Appendix 2.10a)

$$\frac{c}{c_a} = \mathrm{e}^{i} \tag{2.40}$$

where $i = \omega(a-z)/\epsilon_s$, and c_a is the concentration of sediment at a height a above the bed. However, we have already argued that ϵ_s is comparable to ϵ_m and, from Equations (2.9) and (2.10a), we know that

$$\epsilon_m = \kappa^2 z^2 \frac{\mathrm{d}u}{\mathrm{d}z} \;;$$

if we now assume that $\epsilon_s = \epsilon_m$, we find (Appendix 2.10b) that

$$\frac{c}{c_a} = \left[\frac{a(D-z)}{z(D-a)}\right]^{\omega/\kappa u_*} . \tag{2.41}$$

Flume experiments have shown Equation (2.41) to be fairly satisfactory as an expression for the sediment gradient, but it appears that von Kármán's coefficient is *not* constant when applied to sediment mixing. (Indeed, for *turbid* water κ is not even constant for the mixing of momentum.) As a corollary, it follows [Equation (2.10)] that it is invalid to assume $\epsilon_s = \epsilon_m$. Both Leliavsky (1966) and Henderson (1966) provide interesting data on the relationship between these two coefficients. A further unfortunate limitation of Equation (2.41) is that as the limiting depths of flow ($z = 0$ and $z = D$) are approached, the predicted values of sediment concentration tend, unrealistically, to infinity and to zero.

One point that emerges very clearly from Equation (2.41) is that the gradient in sediment concentration is a function of the terminal fall velocity ω and, thus, the particle diameter of the sediment in suspension. For very fine colloidal material $\omega \to 0$ and, from Equation (2.41), $c/c_a \to 1$; that is, the concentration of sediment should be fairly uniform throughout the depth of flow. For coarser material, the gradient in sediment concentration will be much greater. Field evidence (Figure 2.17 substantiates this point although, inevitably, there is usually much scatter about the curve.

Another point that follows from Equation (2.41) is that the capacity f suspended sediment in a river is inversely linked to the water temperature This may be argued as follows. If, for convenience, we assume $a < z$ (th same conclusion emerges if $a > z$), the total suspended load will clearly increase as c/c_a becomes larger (Figure 2.18a). And, if $a < z$, c/c_a is inversely dependent on the fall velocity of the material; in turn, for any given material, the fall velocity is inversely linked to the viscosity. (Ther is more viscous drag acting on the particle and the settling rate is thus reduced; this, of course, is the basis of Stokes' law.) Now, fluids becom more viscous at lower temperatures (Table 2.1); thus the sediment gradient c/c_a and sediment capacity should increase at lower temperatures Naturally, it is unlikely that there will be a close relationship in nature, between temperature and average suspended sediment concentration in streams, because numerous other factors influence the amount of suspended load. However, under rare conditions in which sediment is derived entirely from the bed (and, thus, unaffected by catchment conditions) and discharge is almost constant, observations have shown (Figure 2.18b) that a strong correlation may exist between water temperature and suspended sediment concentration. The particular data of Figure 2.18b relate to a site downstream from a large dam on the Colorado River; hence, discharge is relatively constant and sediment

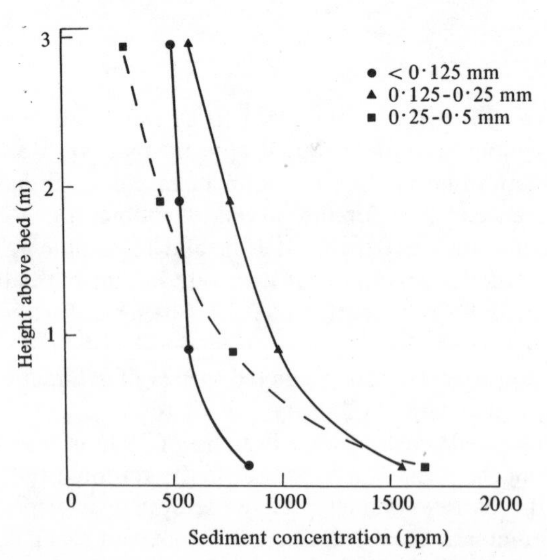

Figure 2.17. Sediment concentration for different particle size classes at different depths, Niobrara River, Nebraska (after Colby and Hembree, 1955).

washed from upland slopes is trapped before reaching the sampling site. Note that, even under these conditions, the fluctuation of sediment concentration with temperature was confined to particles smaller than 0·3 mm in diameter.

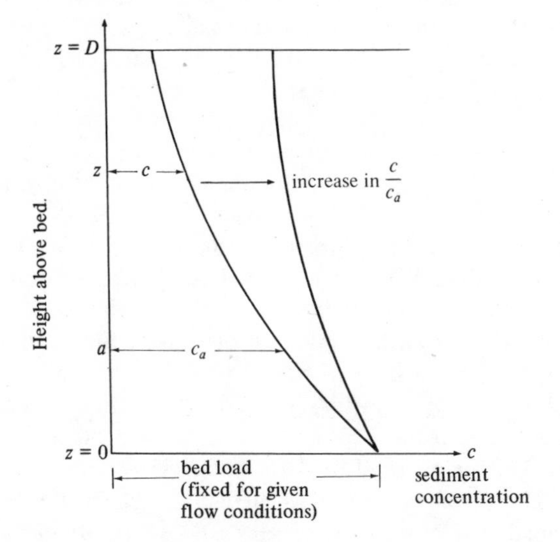

Figure 2.18a. Relation between sediment gradient and total suspended load.

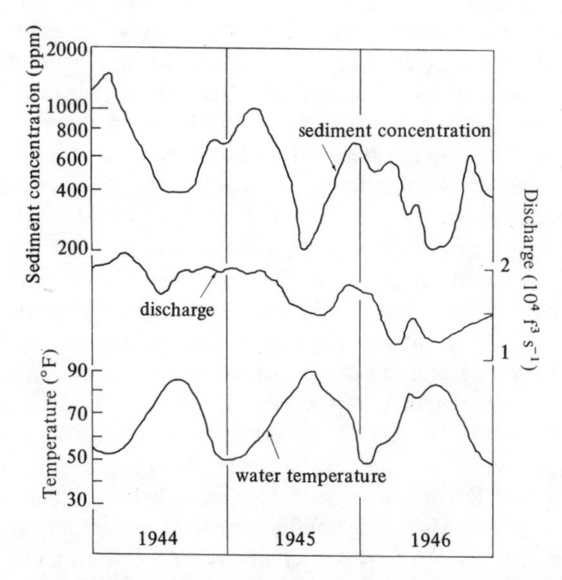

Figure 2.18b. Relationship between suspended sediment concentration and water temperature in the Colorado River (after Lane *et al.*, 1949).

The argument above implicitly assumes that under steady flow conditions the concentration c_a is also constant. Under conditions in which the sediment in suspension is derived entirely from the stream bed this is a plausible assumption. In contrast, in situations where sediment is washed into the channel during the course of a flood, the concentration c_a may increase during the flood, and, in turn, the total suspended sediment discharge may increase without any change in the c/c_a ratio. This point i important, because it should not be thought that, just because the sedime concentration profile of a stream follows a theoretical one based on turbulence theory, the stream is flowing at capacity. A stream is only at capacity for a given size class of material if, under given conditions, c_a is at a maximum for all values of a.

Computations of total suspended sediment discharge may be made usin equations such as (2.40) and (2.41), provided that the concentration at a arbitrary height a above the bed is actually determined in the field. Einstein's (1950) formula for total bed material discharge, that is transpo of debris (either along the bed or in suspension) of size such that it is found on the bed in appreciable quantity, is worthy of particular mentio because a is taken very close to the bed, and it is then possible to predict (rather than measure) c_a from the bed load formula. Unfortunately, for practical rather than theoretical reasons, it is difficult to apply the Einstei method to the wash load (the movement of very fine debris which occurs on the bed in only small quantities) and, to circumvent this drawback, Colby and Hembree (1955) developed the 'modified Einstein procedure' The reader should be quite clear that these formulae are not to be interpreted as predictors of stream capacity; rather, they are procedures for computing actual sediment discharge in relation to stated conditions of channel flow and availability of sediment. Nevertheless, it would seem possible to adapt the Einstein equation by assuming unrestricted availability of sediment of a particular calibre, and predict stream capacity under different flow conditions, subject to the qualifications discussed in the next section.

Much of the discussion above rests on the assumed *random* character o turbulent flow; this is based on the ideas of Prandtl and von Kármán. However, both laboratory and field experiments have shown that the instantaneous cross-currents in turbulent flow are not necessarily random; steady upward currents may exist at some points and steady downward currents at others. In other words, superimposed on the main downstrea flow, a spiralling motion may occur. This has many implications for the mechanics of fluvial erosion. In the first place, it clearly undermines the basis for the Prandtl-von Kármán approach to suspended sediment distribution within a moving fluid. Secondly, it offers a mechanism for periodic scour along the course of stream channels, and using this approac Shen and Komura (1968) have formulated one of the few satisfactory explanations for meander development in initially straight channels.

Unfortunately, a great deal more research is needed before any generalizations can be made, although it is now clear that many of the problems of fluid erosion may be related to the non-random, rather than random, aspects of turbulent flow.

2.4 Threshold erosion conditions

W.M.Davis long ago argued that eventually streams attain equilibrium longitudinal profiles in which, at any point, the energy of the stream is *just* sufficient to transport all the debris transported from upslope, but insufficient to engage in further downcutting at that point, provided that we consider conditions averaged over a period of years. Davis called this condition of dynamic equilibrium, the state of grade. Although Davis' ideas found widespread acceptance among geomorphologists, few of them actually considered testing whether or not 'mature' streams in nature, do, in fact, exist in this condition.

In order to test the Davisian concept, two requirements are needed: (1) a rational formula for predicting the minimum energy conditions of a fluid for transporting the amount of sediment supplied to it; and (2) instrumental techniques for measuring the amount of sediment supplied to a stretch of channel per unit time. Now, the *minimum* energy conditions necessary to transport a *given* amount of sediment can, clearly, be obtained from a formula which predicts the *maximum* amount of sediment that can be transported under *given* energy conditions. Various equations for predicting bed-load capacity have been presented in Section 2.3. Comparable equations for suspended load are rare, but this is unimportant because, for very fine debris, there is no limit to the amount of suspended sediment that can be carried. It would thus appear to be very easy to test the concept of grade. Observations should be made on natural channels over a period of years to determine (a) the rate of bed-load movement, and (b) the prevailing energy conditions of the flow. Using a given bed-load equation, the amount of energy just needed to transport the given amount of sediment could be calculated, and a comparison could be made between actual energy conditions and the predicted minimum or threshold conditions. If the stream is at grade at that point, the two values should be the same.

In practice, this is not an easy task. Firstly, it is difficult to measure the bed-load discharge of natural streams carrying coarse material. Secondly, none of the bed-load formulae has really been proved acceptable for natural channels. In the development of the bed-load formulae, we noted that these are only *semi*-deductive; they involve coefficients and exponents which must be obtained from empirical data. Thus, even if we could measure bed-load accurately, we still face the problem of applying the correct formula. We could, *on the assumption that the natural channels under study are transporting full loads of sediment*, obtain appropriate coefficients and exponents for the bed-load formula under natural conditions. But, it is then meaningless to use this formula to test the concept of grade, at least at the same sites, because, in developing the formula, the condition of grade has already been assumed. Until a fully deductive bed-load equation is developed, this chicken-egg problem must, to varying degrees, frustrate our attempts to test the Davisian concept.

While it is not possible to compare actual and theoretical (graded) stream energy conditions *at a single site*, it is possible to compare *downstream changes* in the actual and theoretical (graded) energy conditions. According to Davis' definition of a graded channel, no extra debris is picked up from the bed at any point and, therefore, the bed-load discharge must be the same at all points along the channel. If we now refer to Einstein's (1942) formula [Equation (2.36c)] and maintain Wq_b constant, where W is the channel width and q_b is bed-load discharge per unit width, we find that the parameter

$$\xi = \frac{W^{1/3}Rs}{d^{1/2}}$$

must also be constant along a graded stretch of stream channel. Now, both R (channel hydraulic radius) and W increase downstream, and d (particle size) decreases downstream. We thus deduce that s must decrease in a downstream direction, that is, the long profile of a graded stream must be concave. The exact nature of the profile will depend on the rate at which R, W, and d change with distance downstream. Many conflicting opinions (see Yatsu, 1955) exist. In view of the interest in fluvial morphometry among geomorphologists, it is surprising that field values of ξ (xi) are so few. Some are shown in Table 2.5. for different floodplain streams on Exmoor and in the southern Pennines. Values of R, W, and s refer to the bankfull state; d is the median particle size. Because bed-load discharge undoubtedly varies from stream to stream, even within the same area, it is not surprising that ξ is not constant; nonetheless, even with this set of values, the variance in ξ is quite restricted.

Finally, note that the ratio $\xi/W^{1/3}d^{1/2}$ is very similar in form to the Shields entrainment function. Indeed, on assuming fully turbulent conditions ($d/\delta_0 \rightarrow \infty$), Figure 2.9 indicates that, at the threshold of bed load movement,

$$\frac{\xi_c}{W^{1/3}d^{1/2}} \approx 0 \cdot 1 , \qquad (2.42$$

where appropriate values for ρ_s, ρ_f, and g have been substituted. Comparison with the data in Table 2.5 shows that actual stream channels have, on average, slightly higher values. This is not surprising. A higher tractive stress (Rs) is necessary to transport a given amount of bed material of size d, than merely to initiate movement of a few particles. The interesting point is that the values are, in fact, so close to the Shields value.

Table 2.5. Bankfull channel characteristics of some floodplain streams on Exmoor and in Pennines.

Site	W (cm)	R (cm)	$s \times 10^2$	d (cm)	ξ	$\frac{\xi}{W^{1/3}d^{1/2}}$
Horner	460	81	2·0	6·8	4·8	0·24
Farley	355	63	2·6	7·1	4·4	0·23
Ilkerton	395	60	1·4	4·2	3·0	0·20
Hoaroak	365	105	1·1	6·0	3·3	0·19
Warren Farm	370	66	0·5	2·5	1·5	0·13
Simonsbath	540	79	1·3	6·6	3·4	0·16
Kinsford	350	60	0·5	3·9	1·3	0·08
Bridgetown	800	108	0·2	4·6	1·0	0·05
Kniveton	245	75	2·2	4·8	4·7	0·34
Turlow	60	55	1·6	2·7	2·0	0·32
Dove	1000	144	0·3	7·0	1·6	0·06
Wye	920	120	1·4	6·9	6·3	0·24
Ecton	1000	155	0·9	4·5	6·5	0·31

Bibliography

Albertson, M. L., Barton, J. R., Simons, D. B., 1960, *Fluid Mechanics for Engineers* (Prentice-Hall, Englewood Cliffs, N.J.).

Bagnold, R. A., 1953, *The Physics of Blown Sand and Desert Dunes* (Methuen, London).

Brown, C. B., 1950, "Sediment transportation" in *Engineering Hydraulics* (Ed. H. Rouse) (John Wiley, New York), pp.769-857.

Brunt, D., 1934, *Physical and Dynamical Meteorology* (Cambridge University Press, Cambridge).

Bruun, P., Lackey, J. B., 1962, "Engineering and biological aspects of sediment transport", *Reviews in Engineering Geology,* **1**, 39-103.

Chepil, W. S., 1945, "Dynamics of wind erosion: 1. Nature of movement of soil by wind", *Soil Sci.,* **60**, 305-320.

Chepil, W. S., 1961, "The use of spheres to measure lift and drag on wind-eroded soil grains", *Soil Sci. Soc. Am. Proc.,* **25**, 343-346.

Chow, Ven Te, 1959, *Open-Channel Hydraulics* (McGraw-Hill, New York).

Colby, B. R., Hembree, C. H., 1955, "Computations of total sediment discharge, Niobrava River, near Cody, Nebraska", *US Geol. Surv. Water Supply Paper* 1357.

Einstein, H. A., 1942, "Formulas for the transportation of bed load", *Trans. Am. Soc. Civil Engrs.,* **107**, 561.

Einstein, H. A., El-Samni, E. A., 1949, "Hydrodynamic forces on a rough wall", *Rev. Mod. Phys.,* **21**, 520-524.

Einstein, H. A., 1950, "The bed-load function for sediment transportation in open channel flow", *U. S. Dep. Agr., Tech. Bull.,* 1026.

Emmett, W. W., 1970, "The hydraulics of overland flow on hillslopes", *U. S. Geol. Surv. Profess. Papers,* 662A.

Helley, E. J., 1969, "Field measurement of the initiation of large bed particle motion in Blue Creek near Klamath, California", *U. S. Geol. Surv. Profess. Papers,* 562G.

Henderson, F. M., 1966, *Open-Channel Flow* (MacMillan, New York), Chapter 10.

Hinze, J., 1959, *Turbulence: An Introduction to Its Mechanism and Theory* (McGraw-Hill, New York).

Hjulström, F., 1935, "Studies of the morphological activity of rivers as illustrated by the River Fyris", *Bull. Geol. Inst. Univ. Uppsala,* **25**, 221–527.

Jeffreys, H., 1929, "On the transport of sediment by streams", *Proc. Cambridge Phil. Soc.,* **25**, 272.

Lane, E. W., Carlson, E. J., Hanson, O. S., 1949, "Low temperature increases sediment transportation in Colorado River", *Civil Eng.,* September, 45-46.

Leliavsky, S., 1966, *An Introduction to Fluvial Hydraulics* (Dover Publications, New York).

Leopold, L. B., Wolman, M. G., Miller, J. P., 1964, *Fluvial Processes in Geomorphology* (Freeman, San Francisco).

Prandtl, L., 1952, *Fluid Dynamics* (Hafner, New York).

Rouse, H. (Ed.), 1959, *Advanced Mechanics of Fluids* (John Wiley, New York).

Rubey, W. W., 1937, "The force required to move particles on a stream bed", *U. S. Geol. Surv. Profess. Papers,* 189E.

Schlichting, H., 1955, *Boundary Layer Theory* (McGraw-Hill, New York).

Sellers, W. D., 1965, *Physical Climatology* (Chicago University Press, Chicago).

Shapiro, A. H., 1961, *Shape and Flow. The Fluid Mechanics of Drag* (Anchor Books, New York).

Shen, H. W., Komura, S., 1968, "Meandering tendencies in straight alluvial channels", *Proc. Am. Soc. Civil Engrs., J. Hydraulics Div.,* HY4, 997-1016.

Shields, A., 1936, "Anwendung der Ähnlichkeitsmechanik und der Turbulenzforschung auf die Geschiebebewegung", *Mitt. preuss. Vers Anst. Wasserb. Schiffbau,* **26**.

Sutton, O. G., 1953, *Micrometeorology* (McGraw-Hill, New York).

Townsend, A. A., 1956, *The Structure of Turbulent Shear Flow* (Chapter 9) (Cambridge University Press, Cambridge).

White, C. M., 1940, "Equilibrium of grains on the bed of a stream", *Proc. Roy. Soc. (London),* **Ser.A, 174**, 322-334.

Yatsu, E., 1955, "On the longitudinal profile of the graded river", *Trans. Am. Geophys. Union,* **36**, 655-663.

Appendix 2

2.1 Definition of some terms relating to fluid flow

streamline: an imaginary line within the flow of a fluid for which, at any point, the tangent to the line indicates *either* the instantaneous *or* the time-average direction of flow at that point.
pathline: the trajectory of *either* an individual *or* an average element of fluid between two points.
steady flow: flow in which, at any *point*, there is no fluctuation in velocity (in magnitude or direction) with respect to time; in steady flow pathlines are identical to streamlines; flow in which $du/dt \neq 0$ is called unsteady flow.
uniform flow: flow in which, at any *time*, there is no variation in velocity with respect to distance along a pathline; from Newton's Second Law of Motion ($P = ma$), this condition ($a = 0$) implies a balance between driving and resisting forces on the fluid; flow in which $du/dx \neq 0$ is called non-uniform flow.
laminar flow: flow in which instantaneous streamlines are parallel; streamline flow is an alternative name.

Note: In any strict definition of steady or uniform flow, some mention must be made of the scale (time or space) of the problem. Steady flow is not incompatible with turbulent flow when it is defined by $d\bar{u}/dt = 0$ in which $\bar{u}$ is the average velocity over a small, but not infinitesimal, time period. Flow may, moreover, be steady over an intermediate time period (for example, hours), but unsteady over a longer period, as shown, for example, by a yearly record.

2.2 Viscosity

As indicated in the main part of the chapter, viscosity can be of two *kinds* (molecular or eddy) depending on the mode of flow. Moreover, for both types, there are two *measures* of viscosity, kinematic and dynamic. *Kinematic* viscosity, used in the main in this chapter, is measured in dimensions of area/time ($L^2\ T^{-1}$), e.g. $cm^2\ s^{-1}$. These dimensions may seem strange, but, as mentioned previously, they may be viewed as a measure of the amount of interference per unit time between parallel layers of fluid in flow. An alternative approach to the kinematic viscosity is as follows. Recall the statement that the fluid shear stress is proportional to the gradient in momentum concentration away from the fluid boundary; that is,

$$\tau = c\frac{d\rho_f u}{dz} = c\rho_f\frac{du}{dz}$$

where c is the constant of proportionality. The dimensions of c are given by

$$ML^{-1}T^{-2} = [c]ML^{-3}T^{-1}$$

or

$$[c] = L^2T^{-1}\ .$$

Viewed this way, the kinematic viscosity is nothing more than the coefficient of proportionality in the statement above.

The alternative measure of viscosity, *dynamic* or *absolute* viscosity, is quite easy to envisage. It is expressed in dimensions of force × time/area ($ML^{-1}T^{-1}$), e.g. lb s ft^{-2}. Consider (Figure A2.1) a cube of fluid being deformed by a shear force T (lb); the amount of deformation (area ABCD in in^2) is a direct function of the magnitude of the applied force and the length of time allowed for deformation, and inversely proportional to the viscosity of the fluid. (Wax, for instance, will deform more slowly than thin oil.) It follows, therefore, that viscosity is, in turn, directly proportional to the force and time necessary to deform a fluid by a given amount (area). An appropriate measure of viscosity is thus given by Tt/a with dimensions of force × time/area; this is the meaning of dynamic viscosity. The two measures of viscosity are, in fact, related; the dynamic viscosity is equal to the kinematic viscosity multiplied by the mass density of the fluid.

2.3 Forms of the exponential fluid flow profile

The basic equation is

$$u = \frac{1}{\kappa} u_* \ln \frac{z}{z_0} \tag{A2.1}$$

where u_*/κ is equivalent to b in Equation (2.4). Sometimes this is also expressed in terms of common logarithms; (A2.1) then becomes

$$u = \frac{2 \cdot 3}{\kappa} u_* \lg \frac{z}{z_0} \ . \tag{A2.2}$$

If κ is taken as $0 \cdot 4$, these two equations reduce to

$$u = 2 \cdot 5 u_* \ln \frac{z}{z_0} \tag{A2.3}$$

and

$$u = 5 \cdot 75 u_* \lg \frac{z}{z_0} \ . \tag{A2.4}$$

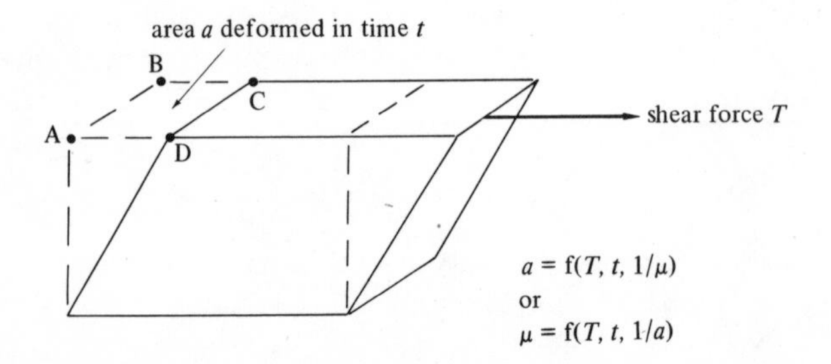

Figure A2.1. The meaning of dynamic viscosity, μ, for molecular interference; μ has the dimensions of force × time/area.

The former is used in Sutton's *Micrometeorology* (p.82) and the latter is employed by Bagnold (1953).

For hydrodynamically *rough* surfaces, the velocity profile is often (e.g. Einstein, 1950, p.8) expressed in the form

$$\frac{u}{u_*} = 5 \cdot 75 \lg \frac{z}{d} + 8 \cdot 5 \qquad \text{(A2.5)}$$

where d is the size of particles on the bed. This is based on the observation [Equation (2.5)] that for rough surfaces $z_0 = d/30$; substituting for d in (A2.5), we obtain

$$\frac{u}{u_*} = 5 \cdot 75 \lg z - 5 \cdot 75 \lg z_0 - 5 \cdot 75 \lg 30 + 8 \cdot 5$$

and, with $5 \cdot 75 \lg 30 = 8 \cdot 5$, we obtain Equation (A2.4).

For hydrodynamically *smooth* surfaces, the velocity profile *above the laminar sublayer* (see Appendix 2.4) is often given by

$$\frac{u}{u_*} = 5 \cdot 75 \lg \frac{u_* z}{\nu} + 5 \cdot 5 \; ; \qquad \text{(A2.6)}$$

this is the form used in Sutton's *Micrometeorology*. It is derived from Equation (2.6):

$$z_0 = \frac{1}{9} \frac{\nu}{u_*} \; .$$

Substituting for z_0 in (A2.4) yields:

$$\frac{u}{u_*} = 5 \cdot 75 \lg z - 5 \cdot 75 \lg \left(\frac{1}{9} \frac{\nu}{u_*} \right)$$

or

$$\frac{u}{u_*} = 5 \cdot 75 \lg \frac{z u_*}{\nu} - 5 \cdot 75 \lg \frac{1}{9} \; ,$$

which reduces to Equation (A.2.6).

2.4 Derivation of z_0 for smooth boundary case

For *laminar* conditions, Equation (2.7) indicates

$$\frac{\mathrm{d}u}{\mathrm{d}z} = \frac{\tau_0}{\rho_f} \frac{1}{\nu} = \frac{u_*^2}{\nu} \; ;$$

on integration, we obtain

$$u = u_*^2 \frac{z}{\nu} = u_* \frac{u_* z}{\nu} \; . \qquad \text{(A2.7)}$$

[Equation (A2.7) combined with (A2.6) gives the complete velocity profile above a smooth boundary.]

The velocity at the top of the laminar sublayer ($z = \delta_0$) is thus

$$u_{\delta_0} = u_* \frac{u_* \delta_0}{\nu} .$$

Experimental evidence [Equation (2.3)] indicates that

$$\frac{(\tau_0/\rho_f)^{1/2} \delta_0}{\nu} = \frac{u_* \delta_0}{\nu} = 11 \cdot 6$$

so that

$$u_{\delta_0} = 11 \cdot 6 u_* .$$

Now, the point $z = \delta_0$ is also part of the *turbulent* boundary layer, so that from Equation (2.4) we have

$$u_{\delta_0} = b \ln \frac{\delta_0}{z_0} .$$

Combining these two expressions for the velocity at $z = \delta_0$ we obtain

$$11 \cdot 6 u_* = b \ln \frac{\delta_0}{z_0}$$

or

$$\ln \frac{\delta_0}{z_0} = \frac{11 \cdot 6 u_*}{b}$$

or

$$\frac{\delta_0}{z_0} = \exp \frac{11 \cdot 6 u_*}{b} .$$

Now, from Equation (2.15),

$$b = \frac{u_*}{\kappa} = \frac{u_*}{0 \cdot 4}$$

whence

$$\frac{\delta_0}{z_0} = \exp \frac{11 \cdot 6 u_* \times 0 \cdot 4}{u_*}$$

and

$$z_0 = \frac{\delta_0}{e^{4 \cdot 64}} = \frac{\delta_0}{103}$$

or, using Equation (2.3),

$$z_0 = \frac{11 \cdot 6}{103} \frac{\nu}{u_*} \approx \frac{\nu}{9 u_*} .$$

2.5 Derivation of the sixth-power law for stream competence

(a) From Equation (2.20),

$$d_c = f(\tau_0)$$

where d_c is the maximum diameter of particle that can be moved under the unit tractive force τ_0. If the weight of a particle is denoted by w, it follows that

$$w_c^{1/3} = f(\tau_0)$$

because

$$w = f(d^3) .$$

Now it can be shown (below) that

$$\tau_0 = f(\bar{u}^2) ,$$

where $\bar{u}$ is mean channel velocity; hence

$$w_c = f(\bar{u}^6) .$$

Thus, if competence is defined as the heaviest particle that can be transported by the stream, it is a function of the sixth power of mean channel velocity.

(b) Consider the flow of water through a circular pipe over a distance L; the amount of energy lost (to heat) per unit volume of fluid is indicated by the fall in the piezometric head h_f (see Appendix 2.7) as shown in Figure A2.2. In units of pressure, this loss is equal to $h_f(\rho_f g)$. The total energy loss over a cross-sectional area A in flowing over a distance L is thus given by

$$P_1 = A h_f \rho_f g$$

with dimensions of force. This energy loss is due to boundary resistance; the magnitude of this resistance over a distance L is equal to

$$P_2 = \bar{\tau}_0 P_w L$$

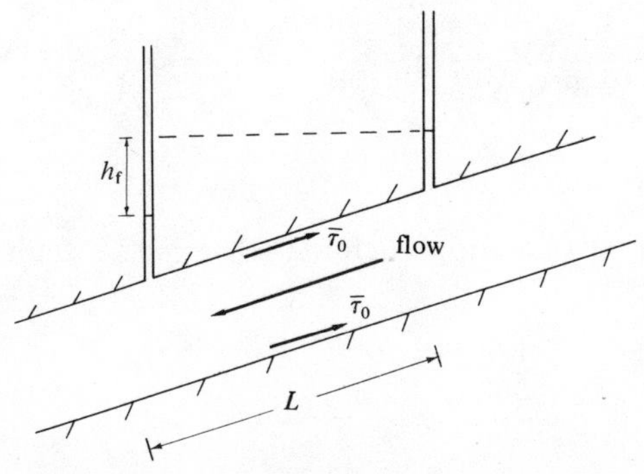

Figure A2.2. Energy loss in fluid flow through a pipe.

where P_w is the wetted perimeter. The forces P_1 and P_2 are merely alternative expressions for the energy dissipation; accordingly, we have

$$\overline{\tau}_0 P_w L = A h_f \rho_f g$$

or

$$\frac{\overline{\tau}_0}{\rho_f g} = \frac{A}{P_w}\frac{1}{L}h_f \; . \qquad \text{(A2.8)}$$

Commonly both sides of this equation are made dimensionless by dividing by the velocity head $(\overline{u}^2/2g)$ (see Appendix 2.7) where $\overline{u}$ is the average velocity through the cross-section. This yields

$$\frac{\overline{\tau}_0/\rho_f g}{\overline{u}^2/2g} = \frac{R h_f/L}{\overline{u}^2/2g} \; ;$$

or

$$\frac{4\overline{\tau}_0/\rho_f g}{\overline{u}^2/2g} = \frac{D h_f/L}{\overline{u}^2/2g} \; , \qquad \text{(A2.9)}$$

if we substitute $R = \frac{1}{4}\pi D^2/\pi D = \frac{1}{4}D$ for circular pipes. The right-hand side of this equation is defined as the (dimensionless) Darcy-Weisbach resistance coefficient; it is denoted by f. Substituting this coefficient, we obtain

$$\overline{\tau}_0 = \tfrac{1}{8} f \overline{u}^2 \rho_f \; , \qquad \text{(A2.10)}$$

which indicates that the boundary shear stress is a function of the square of the mean pipe velocity. The same equation may be applied to open channel flow provided that $4R$ (rather than R) is used in the determination of the friction coefficient.

2.6 Theoretical limit to the square root relationship between particle size and threshold shear velocity [Equation (2.21)]

From Equation (2.3) we find the thickness of the laminar sublayer as

$$\delta_0 = 11{\cdot}6\nu\frac{1}{u_*} \; ,$$

which may be rearranged to give

$$u_* = 11{\cdot}6\nu\frac{1}{\delta_0} \; .$$

From Equation (2.22), the maximum particle size that can be transported by fluid flow is

$$d_c = \frac{\rho_f}{\rho'}\frac{u_*^2}{gA^2}$$

or, substituting for u_*,

$$d_c = \frac{\rho_f}{\rho'} \frac{(11 \cdot 6\nu)^2}{\delta_0^2} \frac{1}{gA^2} . \tag{A2.11}$$

When $d = \delta_0$, particles are just submerged in the laminar sublayer and, at the threshold of particle movement, this critical maximum particle size is obtained by substituting $\delta_0 = d_c$ in (A2.11)

$$d_c^3 = \frac{\rho_f}{\rho'} \frac{(11 \cdot 6\nu)^2}{gA^2} \qquad \text{[Equation (2.23)]} .$$

For air, on taking $\nu = 1 \cdot 35 \times 10^{-1}$ cm^2 s^{-1} (at 40°F), $A^2 = 0 \cdot 01$ (from Bagnold's experiments) and $\rho_f/\rho' = 5 \times 10^{-4}$, Equation (2.23) reduces to $d_c = 0 \cdot 05$ cm $= 0 \cdot 5$ mm.

For water, on taking $\nu = 1 \cdot 35 \times 10^{-2}$ cm^2 s^{-1} (at 48°F), $A^2 = 0 \cdot 04$ (from Shields' experiments) and $\rho_f/\rho' = 6 \cdot 2 \times 10^{-1}$, Equation (2.23) reduces to $d_c = 0 \cdot 7$ mm.

For air, the theoretical figure $d = 0 \cdot 5$ mm compares with an actual value of $d = 0 \cdot 2$ mm at which the square root relationship between particle size and critical shear velocity (or maximum particle size and actual shear velocity) begins to break down (Figure 2.10). In the case of water, it is not possible to compare the theoretical value with Hjulström's curve because data are given only for mean channel velocity ($\overline{u}$), and not shear velocity. There is, however, a close relationship between shear velocity and mean channel velocity, as indicated in Equation (A2.10):

$$u_*^2 = \tfrac{1}{8} f \overline{u}^2$$

or

$$u_* = \frac{\overline{u}}{C/\sqrt{g}}$$

where C $(= \sqrt{8g/f})$ is the well-known Chezy coefficient. Note that both of these resistance coefficients are related to the much-used Manning n measure of roughness; it is easily shown that

$$C \approx 1 \cdot 5 \frac{R^{1/6}}{n}$$

where R (in feet) is the hydraulic radius. Typical values of n for natural open channels are about $0 \cdot 02$–$0 \cdot 06$ ft$^{1/6}$; an appropriate value for $C/\sqrt{g}$ (dimensionless) would therefore be about 12. Hjulstrom's values of mean velocity in Figure 2.10 have thus been reduced by one-twelfth when plotted as u_* in Figure 2.11a. Close agreement is shown between the predicted value ($0 \cdot 7$ mm) and the lower limit of the square root plot ($0 \cdot 8$ mm).

2.7 Bernoulli's equation for steady flow of incompressible fluids

The law of the conservation of energy states that energy is neither created nor destroyed. Applied to open channel flow, this law indicates that, for steady flow, the sum of four components of energy must be constant at different points in that flow:

$$E_k + E_p + E_e + E_h = \text{constant} ,$$

where E_k denotes kinetic energy, E_p is pressure energy, E_e is potential energy due to elevation, and E_h is heat. The amount of energy associated with each component may be expressed in dimensions of force × length (energy) or in dimensions of length (energy per unit weight, or head):

	Energy	Energy per unit weight (head)		
kinetic energy	$\frac{1}{2}mu^2$	$\frac{u^2}{2g}$	velocity head	
pressure energy	$\frac{m}{\rho_f}p$	$\frac{p}{\rho_f g}$	pressure head	piezometric head
elevation energy	mgh	h	elevation head	

In open channel flow, heat energy is created by boundary resistance; the sum $(E_k + E_p + E_e)$ is therefore decreased in moving downstream between two points according to the following equation:

$$\frac{u_1^2}{2g} + \frac{p_1}{\rho_f g} + h_1 = \frac{u_2^2}{2g} + \frac{p_2}{\rho_f g} + h_2 + \frac{\Delta E_h}{mg} . \qquad \text{(A2.12)}$$

Note that for uniform flow $u_1 = u_2$. Moreover, for open channel flow $p_1 = p_2$ (at the surface $p_1 = p_2 = 0$). Under these conditions, Equation (A2.12) simplifies to

$$h_1 - h_2 = \frac{\Delta E_h}{mg}$$

where $\Delta E_h/mg$ also represents h_f (introduced in Appendix 2.5). Now $h_1 - h_2$ is the vertical drop in the elevation of the water surface; it follows, therefore, that the slope of the water surface is a measure of the amount of energy 'loss' per unit distance downstream under conditions of steady, uniform flow.

Note further that over *very* small distances $E_h \rightarrow 0$; consequently changes in any of the components of energy produce changes in one or more of the others. Jeffreys used this in the derivation of the lift force acting on particles on the stream bed: the increase in velocity with height above the bed produces a decrease in pressure away from the bed. A formal proof of Equation (A2.12) will be found in any of the standard textbooks of fluid mechanics, e.g. Albertson *et al.* (1960), p.120.

2.8 The drag coefficient of a particle submerged in fluid flow

The drag force on a submerged particle may be due to two components, *skin friction* drag and *form* (or pressure) drag, illustrated in Figure A2.3.

Skin friction drag is the shear force imposed by a moving fluid on a boundary surface parallel to the direction of fluid flow. By analogy with the flow of fluid past the walls of a pipe (A2.10), we may write

$$\bar{\tau}_0 = c\frac{\rho_f}{2}\bar{u}^2$$

where c is a proportionality constant. The force (rather than stress) on a submerged object of surface area a is

$$F_s = c_s a\frac{\rho_f}{2}\bar{u}^2 \qquad \text{(A2.13)}$$

where c_s is another proportionality coefficient and u (= f[$\bar{u}$]) is the velocity away from the boundary layer of the object.

Now consider the form drag on a particle; this is due to the difference in fluid pressure on the upstream and downstream faces of a particle. From the Bernoulli equation (A2.12), it may be argued that the pressure on the upstream face (in excess of the mean pressure) is equal to $\frac{1}{2}\rho_f u^2$ (Figure A2.3), where u is the velocity at a point, upstream from the particle, unaffected by the disturbance. Similarly, the pressure on the downstream face may be expected to be a function of $\frac{1}{2}\rho_f u^2$ although considerably influenced by the shape of the particle. We may thus express the *net* pressure drag in the form:

$$F_f = c_f a\frac{\rho_f}{2}u^2 \qquad \text{(A2.14)}$$

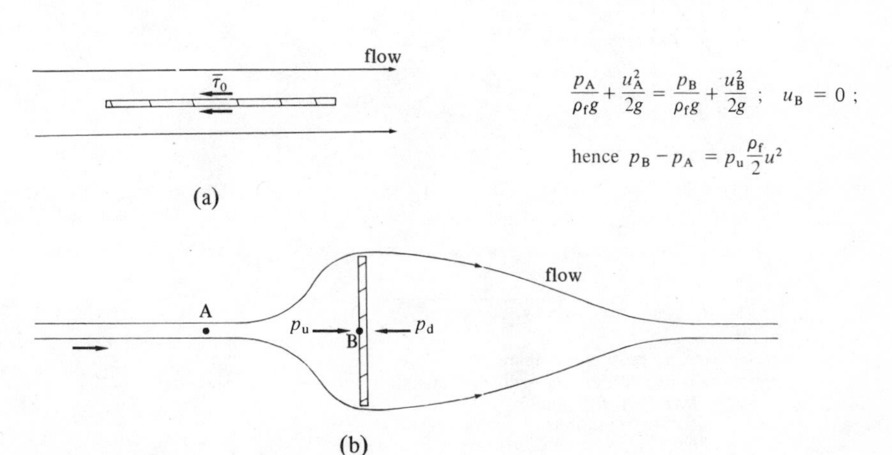

Figure A2.3. The two types of drag in fluid flow past an object: (a) skin friction drag (τ_0); (b) form drag ($p_u - p_d$).

where c_f is strongly influenced by the bluntness of the particle relative to the flow. Both skin friction drag and form drag occur on all particles submerged in a flowing fluid and, commonly, Equations (A.2.13) and (A2.14) are combined to give

$$F_D = C_D a \frac{\rho_f}{2} u^2 , \qquad \text{(A2.15)}$$

where C_D is called the general drag coefficient. Note that there is no unique C_D value for a given particle shape; as might be expected it varies with the mode of flow and, thus, the Reynolds number.

2.9 Bagnold's expression for the rate of bed-load movement in air

Given

$$q_b = \frac{4B}{3} \frac{\rho_f}{g} u_*^3 , \qquad \text{[Equation (2.31)]}$$

the problem is to substitute for u_* in terms of actual wind velocities at given heights above the ground. From Equations (2.4), (2.15), and (2.16)

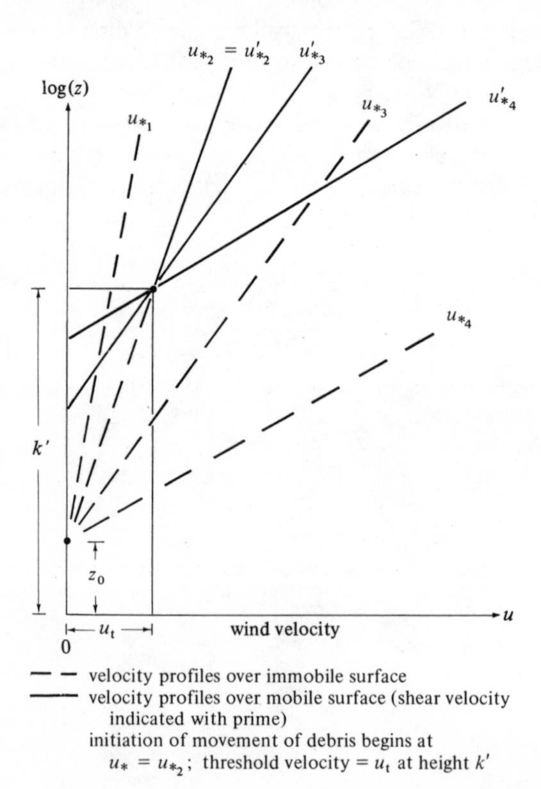

Figure A2.4. Contrast in velocity profiles for mobile and immobile surfaces (based on Bagnold, 1953).

we have

$$u_* = b\kappa = \frac{\kappa u}{\ln(z/z_0)} = \frac{\kappa}{2\cdot 3}\frac{u}{\lg(z/z_0)} \,,$$

where u is the velocity at height z above the ground. Substituting $\kappa = 0\cdot 4$ we obtain

$$u_* = \frac{0\cdot 174u}{\lg(z/z_0)} \,.$$

Actually this expression for the shear velocity is valid only for conditions of *no* bed-load movement; if bed load is being moved, the change in the velocity profile (Figure A2.4) necessitates an adjustment in the formula above to:

$$u_* = \frac{0\cdot 174(u-u_t)}{\lg(z/k')}$$

where k' is analogous, but not equivalent, to z_0, and $(u-u_t)$ is the *excess* velocity at height z (above the velocity at k'). This is self-evident from Figure A2.4.

Equation (2.31) may now be expressed as follows:

$$q_b = \frac{4B}{3}\frac{\rho_f}{g}\frac{0\cdot 174^3(u-u_t)^3}{\lg^3(z/k')} \,. \tag{A2.16}$$

For a given type of sand, $\lg(z/k')$ is constant. Thus, for a fixed reference elevation z, (A2.16) reduces to

$$q_b = \alpha\frac{4B}{3}\frac{\rho_f}{g}(u-u_t)^3 \,, \tag{A2.17}$$

where $\alpha = 0\cdot 174^3/\lg^3(z/k')$.

2.10 Gradient of suspended sediment concentration with height above stream bed

Derivation of Equation (2.40):

$$\frac{c}{c_a} = e^i \,, \qquad i = \frac{\omega}{\epsilon_s}(a-z) \,.$$

From Equation (2.31) we have

$$\epsilon_s\frac{dc}{dz} = -c\omega$$

or

$$\int \epsilon_s\frac{dc}{c} = -\int \omega\, dz$$

or

$$\epsilon_s \ln(c) = -\omega z + A \,. \tag{A2.18}$$

If, for $z = a$, we define $c = c_a$, we have

$$\epsilon_s \ln c_a = -\omega a + A$$

or

$$A = \epsilon_s \ln c_a - \omega a \, .$$

Substituting for A in (A2.18) we obtain

$$\epsilon_s \ln c = -\omega z + \epsilon_s \ln c_a + \omega a$$

or

$$\epsilon_s \ln \frac{c}{c_a} = \omega(a-z)$$

or

$$\frac{c}{c_a} = \exp\left[\frac{\omega}{\epsilon_s}(a-z)\right] .$$

Derivation of Equation (2.41):

$$\frac{c}{c_a} = \left(\frac{D-z}{z}\frac{a}{D-a}\right)^{\omega/u_*\kappa} .$$

From Equation (2.8) we have

$$\epsilon_m = \frac{\tau}{\rho_f} \bigg/ \frac{du}{dz}$$

or, because $\tau/\tau_0 = (D-z)/D$ [see Figure 3.20 and Equation (3.21)],

$$\epsilon_m = \frac{\tau_0}{\rho_f}\frac{D-z}{D} \bigg/ \frac{du}{dz} .$$

From Equation (2.14) we have

$$\frac{du}{dz} = \frac{1}{\kappa z}\sqrt{\frac{\tau_0}{\rho_f}} ;$$

substituting for du/dz in the expression for ϵ_m we obtain

$$\epsilon_m = \frac{\tau_0}{\rho_f}\frac{D-z}{D}\kappa z\sqrt{\frac{\rho_f}{\tau_0}} = u_*\kappa z\frac{D-z}{D} .$$

Substituting $\epsilon_s = \epsilon_m$ in Equation (2.39), we obtain

$$u_*\kappa z\frac{D-z}{D}\frac{dc}{dz} = -c\omega$$

or

$$u_*\kappa\int\frac{dc}{c} = -\omega D\int\frac{dz}{z(D-z)}$$

or

$$u_* \kappa \ln(c) = -\omega D\left(-\frac{1}{D}\ln\frac{D-z}{z}\right) + A\,. \tag{A2.19}$$

If, for $z = a$, we define $c = c_a$, we then have

$$u_* \kappa \ln c_a = \omega \ln\frac{D-a}{a} + A$$

or

$$A = u_* \kappa \ln c_a - \omega \ln\frac{D-a}{a}\,.$$

Substituting for A in (A2.19), we find

$$u_* \kappa \ln c = \omega \ln\frac{D-z}{z} + u_* \kappa \ln c_a - \omega \ln\frac{D-a}{a}$$

or

$$u_* \kappa \ln\frac{c}{c_a} = \omega \ln\left(\frac{D-z}{z}\,\frac{a}{D-a}\right)$$

or

$$\frac{c}{c_a} = \left(\frac{D-z}{z}\,\frac{a}{D-a}\right)^{\omega/u_* \kappa}\,.$$

3

Stress–strain–strength interrelationships

Glossary of symbols

- a fraction of shear surface occupied by inter-granular contacts ($= A_s/A$) [93]
- c cohesion (as a stress; general symbol) [73]
- c' cohesion (as a stress; defined in terms of effective stresses) [74]
- e void ratio [76]
- f( ) function of [73]
- g acceleration due to gravity [87]
- i slope angle [81]
- k coefficient of at-rest earth pressure [77]; Hookean modulus [66]
- k_A coefficient of active earth pressure [78]
- k_P coefficient of passive earth pressure [78]
- l, m, n direction cosines (as used in Chapter 1) [96]
- l length [65]
- δl incremental length [65]
- l_0 initial length [65]
- n exponent in flow law [87]
- p depth to potential failure plane [88]
- r radius [78]
- s shear strength (as a stress) (subscripts p and r denote peak and residual strength respectively) [72]
- s_d shear resistance [68]
- u pore pressure [75]; velocity in direction of flow [88]
- δu_i, δu_{ii} incremental pore pressures [75]
- u_b basal velocity [88]
- u_s surface velocity [88]
- w moisture content (by weight) [72]
- x, y, z Cartesian distance coordinates (z is *either* vertical *or* normal to average flow) [77]

- A area of shear surface [93]
- A_s area of contact between particles on shear surface [93]
- $\bar{A}$ coefficient for (simple shear) flow law [87]
- A' coefficient for (uniaxial compression) flow law [90]
- B coefficient [88]
- C constant of integration [88]
- D depth of fluid flow [87]
- S shear strength (as a force) [73]
- W weight of column of soil [88]
- Z_0 thickness of tension zone [80]

- α angle between failure plane and major principal plane [72]
- β angle relative to horizontal [79]

γ bulk unit weight ($= \rho g$) [77]; shear strain [84]
$\dot{\gamma}$ shear strain rate ($= d\gamma/dt$) [87]
γ_{oct} octahedral shear strain [86]
$\dot{\gamma}_{oct}$ octahedral shear strain rate [86]
ϵ linear strain [65]
$\dot{\epsilon}$ linear strain rate ($= d\epsilon/dt$) [67]
$\epsilon_1, \epsilon_2, \epsilon_3$ principal strains [86]
η dynamic eddy viscosity [89]
θ angle [77]
μ dynamic molecular viscosity [89]; coefficient of friction [68]
ρ bulk mass density [87]
σ normal stress [67]
σ' effective normal stresses [74]
$\delta\sigma_i, \delta\sigma_{ii}$ incremental normal stress [75]
σ_1, σ_3 major and minor principal stresses [72]
σ_{oct} octahedral normal stress [86]
σ_0 yield stress in compression [66]
σ_i intergranular stress [93]
σ_j arbitrary value of σ [74]
σ_m mean normal stress [86]
τ shear stress [72]
τ_{oct} octahedral shear stress [86]
τ_y yield stress in shear [90]
ϕ angle of internal friction ($= \tan^{-1} \mu$) (general symbol) [71]
ϕ' angle of internal friction (defined in terms of effective stresses) [74]

3.1 Strength

We shall begin this section by discussing the stress-strain behaviour of idealized materials; this should provide us with the basic concepts for investigating the behaviour of natural materials. The term *strain* refers to the amount of deformation of a body. For the moment, let us confine our attention to the stress-strain relationship in a triaxial compression test (Chapter 1). The amount of strain in the specimen is defined as

$$\epsilon = \frac{\delta l}{l_0} ,$$

where l_0 is the initial length of the specimen, $(l_0 - \delta l)$ is the length after compression, and ϵ (epsilon) is the strain expressed as a fraction. Strictly speaking, this should be described as conventional strain (Appendix 5.1), but this is not important at the present stage.

Three types of idealized behaviour are shown, separately and in combination, in Figure 3.1:

1. *Elastic* behaviour involves (a) a *finite* amount of strain associated with a particular stress, and (b) fully-recoverable deformation, that is, the amount of strain produced by a stress increase from σ to $(\sigma + \delta\sigma)$ is also

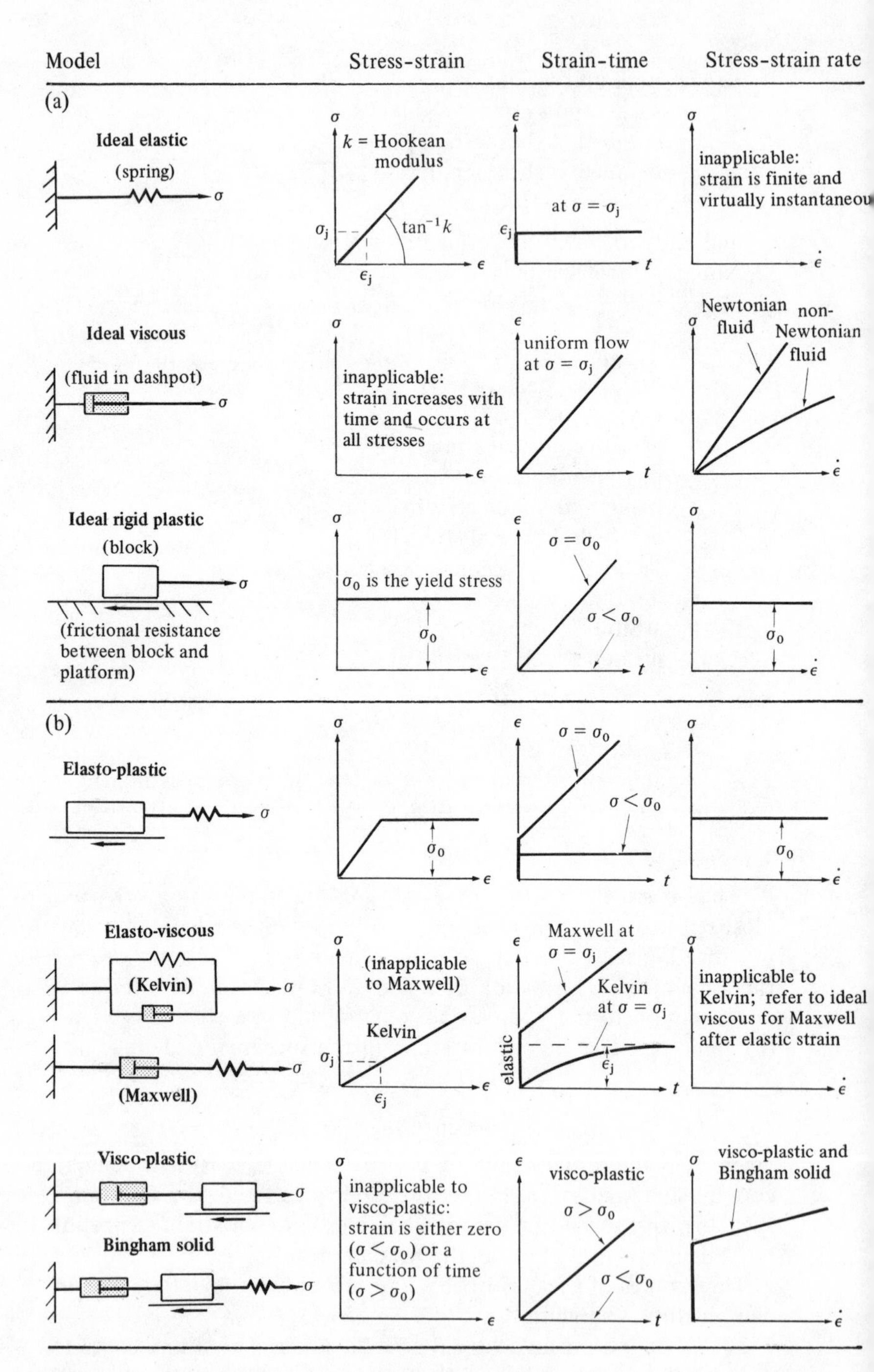

Figure 3.1. Rheological behaviour of (a) ideal materials (elastic, viscous, rigid plastic) and (b) composite materials (elasto-plastic, elasto-viscous, visco-plastic, Bingham solid).

produced by a relaxation of stress from $(\sigma + \delta\sigma)$ to σ. The Hookean model describes one particular type of elastic behaviour in which strain is a *linear* function of stress.

2. *Viscous* materials, in contrast, strain *indefinitely* if a stress is applied, and, in this case, it is the *rate* of strain ($\dot{\epsilon} = d\epsilon/dt$) which is a function of the stress. A fluid, for which the relationship between rate of strain and stress is linear, is defined as a Newtonian fluid.
3. Note that both elastic and viscous materials will deform under any stress irrespective of its magnitude. *Rigid plastic* materials, however, will not deform until a threshold (yield) stress is attained. Such materials are termed ideally *plastic* if it is impossible to increase the applied stress beyond this critical value. Materials which are capable of supporting stresses in excess of the yield value are referred to as *strain-hardening* materials. In contrast, those materials which, after the yield stress has been attained, experience a decrease in the maximum stress which can be supported, are termed *strain-softening.*

Elasto-plastic behaviour, often referred to as the St. Venant model, is a simple combination of idealized elastic and plastic behaviour. Many models of *elasto-viscous* behaviour have been developed; the Maxwell and the Kelvin models, shown in Figure 3.1b, represent two extreme cases. *Visco-plastic* behaviour is sometimes referred to as the Bingham model, although this latter term should be reserved strictly for visco-St. Venant behaviour. The stress–strain–time behaviour of natural materials is more complex than any of these idealized forms, but, to varying degrees, may combine all three basic patterns.

There are two main approaches to the measurement of the strength of materials: stress-controlled and strain-controlled. The former was discussed in connection with the triaxial test in Chapter 1: the axial stress acting on the specimen is increased during the test and the amount of strain associated with each stress increment is noted until, at a critical stress, the specimen deforms continuously as long as this stress is maintained. In the strain-controlled test, the specimen is strained (compressed) at a constant rate, and the amount of resistance to deformation (measured on a proving ring, or similar apparatus, as indicated in Figure 3.2b) is noted at different times during the test. Irrespective of the mode of testing, it is engineering convention to plot stress on the vertical axis and strain horizontally. A typical stress–strain plot for a natural soil specimen is shown in Figure 3.2a. Three stages can be distinguished in the test. Initially the magnitude of the resistance to deformation offered by the specimen increases with time, as the strain increases, until a peak is attained. After the peak there is a gradual reduction in the amount of resistance until, finally, a constant residual value emerges.

It should be appreciated therefore that the magnitude of resistance to deformation is not an instantaneous property, but is developed by the process of strain. *Strength*, on the other hand, is usually regarded as an intrinsic property of a material. It refers to the maximum resistance which can be mustered by a material under given conditions. To some workers, the term 'strength' should be applied only to a material which possesses a plastic component in its rheological behaviour. For ideal elastic materials, resistance is a function of the amount of strain (until the elastic limit, when rupture occurs), and for ideal viscous materials, the resistance to deformation is dependent upon the rate of strain. Only in the case of plastic or pseudo-plastic materials is it possible to obtain a constant resistance over a range of strains and strain rates. Indeed, as noted previously, it is only for ideal rigid plastic materials that a constant resistance is developed for *all* strain and strain rates, but it is meaningful to use the term strength for other types of plastic materials. The plot of Figure 3.2 shows that the resistance of the particular material under study is independent of strain in two parts of the stress-strain curve. The first time is, briefly, during the period when peak resistance is attained; the resistance at this time is accordingly referred to as *peak* strength. The second time is during the development of large strains; the resistance here is termed *residual* strength.

The reasons for the decrease in strength between the peak and residual values in many natural materials are not fully understood. It is tempting to point to the analogous and well-known reduction in the friction coefficient μ (mu) between static and sliding conditions, but this is probably only a minor element. More important is the tendency of platy

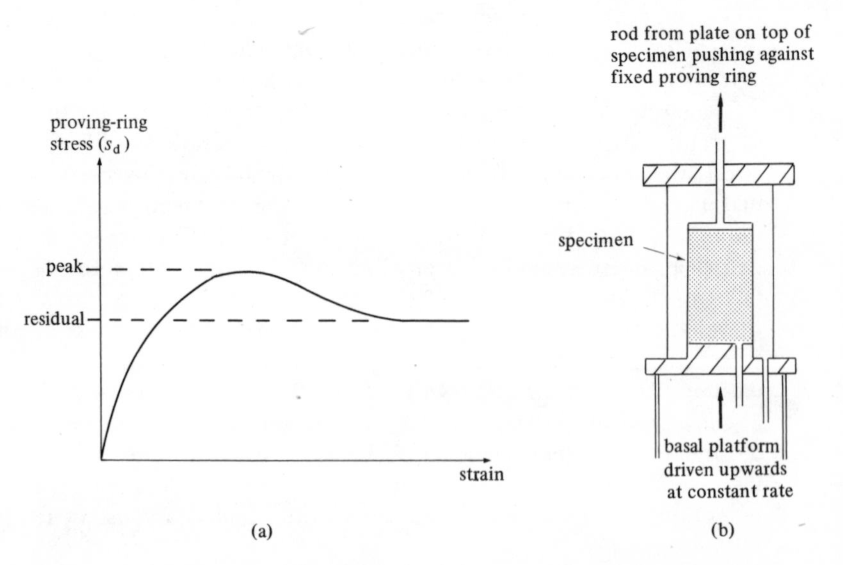

Figure 3.2. Determination of strength using the strain-controlled approach: (a) stress-strain relationship; (b) strain-controlled triaxial compression procedure.

particles to orientate themselves during the process of shear so that interlocking among particles is reduced. Other factors are currently being examined by engineers.

Some materials which exhibit strain-softening, as indicated in Figure 3.2, possess strength values which are entirely independent of the *rate* of strain. Sandy soils often behave in this way, but many clays do not. Stress-controlled tests for typical sandy and clay soil samples are contrasted in Figure 3.3. The stress-strain curve for the sandy soil is similar to the first part of the strain-controlled curve of Figure 3.2a. Each increment of stress produces a corresponding finite amount of strain; the amount of incremental strain per stress increment increases as the overall compressive stress increases. Eventually a sufficiently large stress is built up that the strain becomes infinite, increasing with the duration of the applied stress. This process of continuing deformation is commonly called *yielding* and the stress acting at the onset of yielding is frequently referred to as the *yield stress*. Note that in this particular case, the yield stress is the same as the peak strength of the sample. Note also that the onset of yielding in this particular case is also referred to as *failure* of the sample. Failure is, unfortunately, a rather nebulous term. Some workers define failure as a very large, but fixed, amount of strain, and if we accept this definition we

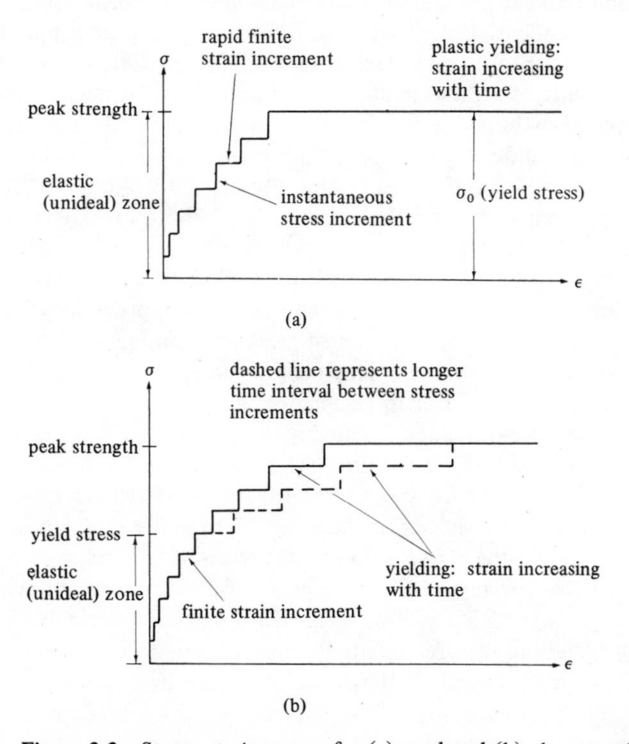

Figure 3.3. Stress-strain curves for (a) sand and (b) clay samples.

shall observe that, for some materials, the stress at failure is equal to neither the yield stress nor the peak strength. However, for the sand sample shown in Figure 3.3, all three stresses are equal.

Now let us examine the stress-strain curve for the clay specimen in Figure 3.3. Note that the onset of yielding occurs at a stress less than th peak strength. In the case of the sandy specimen, the initial rate of yielding is so high that it is impossible to build up the compressive stress beyond the yield value. In the case of many clays, in contrast, the initial rate of yielding is slow, and it is possible to increase the compressive stres beyond the value necessary to initiate yielding; the effect is to increase the rate of yielding. Additional increments of stress increase the yielding rate still further until eventually it becomes so rapid that no additional stress can be supported. The yield strength (the yield stress) and the pea strength are thus not necessarily the same for many clays, and this raises some doubt as to the meaning of the term strength. There is also some confusion over the definition of the term failure. Large strains will occur at stresses below peak strength, provided that the stress is applied for a long enough period of time.

These problems are further complicated by the existence of different types of yielding, or, more accurately, different types of *creep*. The term creep is usually associated with slow *time-dependent strain* in solid materials. Neither ideal elastic materials nor ideal rigid plastic materials exhibit creep. Under an applied stress, ideal elastic materials achieve a small, finite amount of strain virtually instantaneously; under an applied stress, ideal rigid plastic materials either do not strain at all or strain so quickly, as in a landslide, that the strain process is, again, virtually instantaneous. Creep is, therefore, commonly attributed (e.g. Wu, 1966, p.92) to viscous components of the stress-strain behavioral patterns of solids.

Three basic types of creep are shown in Figure 3.4. Attenuating creep (a) occurs at stresses below the yield stress, as defined previously; the strain rate decreases in the later stages of strain, becoming asymptotic towards a particular strain value. This type of creep is essentially elastic in character, modified by a Kelvin viscous component as indicated in Figure 3.1b. The development of a steady creep rate (b) is typical of many natural materials. If we confine our attention to the linear part of the strain-time curve and plot strain-rate for different stress values, we obtain curves of the type shown in Figure 3.5. In some cases the curves become asymptotic to the ϵ axis at a certain stress value and, as in Figure 3.3b, this can be regarded as the peak strength of the material. In contrast, the curves for Boston Blue Clay (Taylor, 1948) and Glen's (195 data for polycrystalline ice are essentially visco-plastic (Figure 3.1b) and c not show a final peak strength. If we define failure in terms of a certain *amount* of strain, it is clear, therefore, that this can occur at various stres values, depending on the duration of the application of the stress. This remark is even more pertinent for the third type of creep (accelerating creep) shown in Figure 3.4; the peak strength value of a material susceptible to accelerating creep (progressive failure) is thus a completely irrelevant parameter in long-term stability problems. Note that many materials exhibit all three types of creep according to the magnitude of the applied stress (attenuating creep at low stresses, accelerating creep at high stresses) but there are also exceptions to this generalization.

Strain rate is only one factor affecting the strength of natural materials in a triaxial compression test. For specimens that are, in other respects, identical, the strength value will also depend upon the magnitude of the all-round cell pressure, the pore pressure inside the specimen, and the initial density of the specimen. These are discussed, in turn, below. First, however, we need to introduce the term *shear strength* (as distinct from compressive strength) and outline the *direct shear* apparatus as an alternative to the compression apparatus. By then, we shall be able to interpret compression test data in terms of *shear* strength parameters.

In Section 3.2, we shall show that specimens under stress fail (in the two-dimensional case) along surfaces which make an angle $\alpha = \pi/4 + \phi/2$ with the direction of the major principal plane. The symbol ϕ (phi) denotes the angle of internal friction of the material; for the moment it is sufficient to regard $\phi = \tan^{-1}\mu$, μ being the coefficient of friction. At the time of incipient failure in the triaxial apparatus, the shear stresses acting

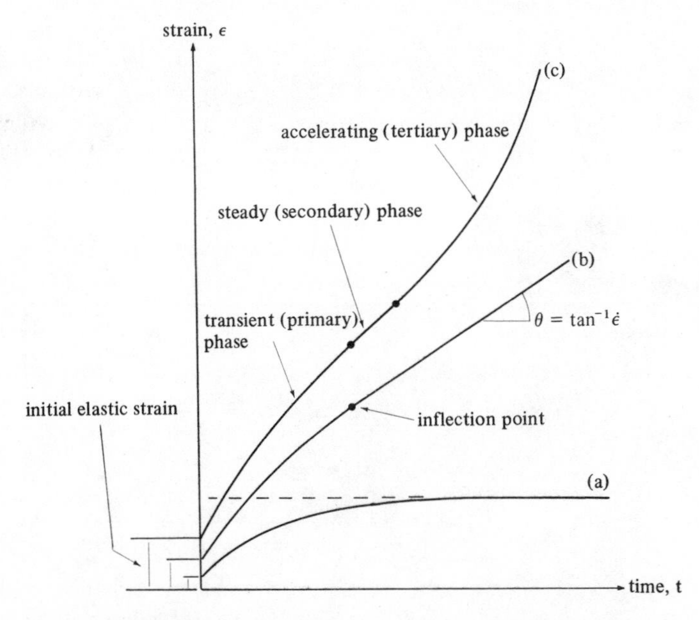

Figure 3.4. Types of creep: (a) attenuating creep; (b) development of steady creep rate; (c) development of accelerating creep (progressive failure). Each test undertaken with constant compressive force.

along these failure surfaces are just balanced by the resistance against shearing offered by the specimen (Figure 1.2); this resistance is called shear resistance. At the instant of failure we therefore have

$$b = \alpha = \pi/4 + \phi/2$$

and, from Equation (1.6),

$$\tau = \tfrac{1}{2}(\sigma_1 - \sigma_3)\sin(\pi/2 + \phi)$$

or, denoting the maximum shear resistance, that is, the shear strength (as a stress rather than a force) by s,

$$s = \tau = \tfrac{1}{2}(\sigma_1 - \sigma_3)\cos\phi \,. \tag{3.1}$$

Similarly, from Equation (1.5), we can determine the magnitude of the normal stress on the failure plane; we obtain

$$\sigma = \tfrac{1}{2}(\sigma_1 + \sigma_3) - \tfrac{1}{2}(\sigma_1 - \sigma_3)\sin\phi \,. \tag{3.2}$$

The shear and normal stresses acting on the failure surfaces in a triaxial test are thus determined indirectly from the cell pressure and the applied vertical stress. In the direct shear test, these two parameters are measured

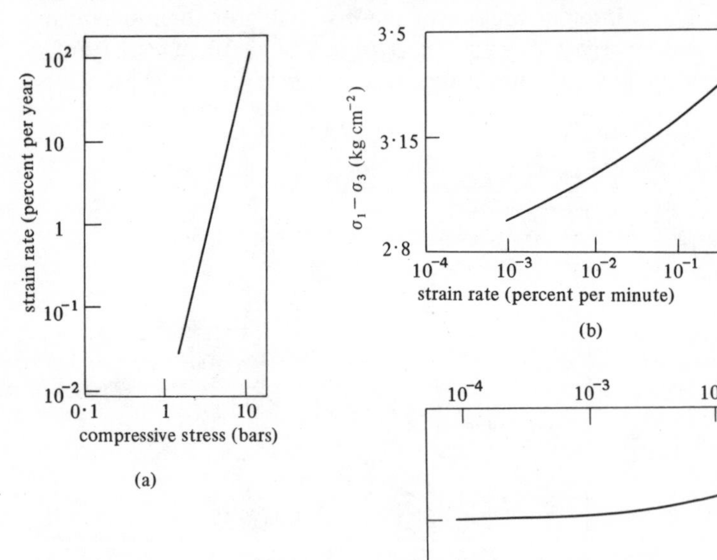

Figure 3.5. Stress-strain rate curves for various natural materials: (a) polycrystalline ice at −2·5°C to −0·8°C (after Glen, 1952); (b) remoulded Boston Blue Clay at 19% water content (after Taylor, 1948); (c) weathered Edale Shales—direct shear tests.

directly. The direct shear apparatus is shown in Figure 3.6. The specimen is contained in a box split into upper and lower frames. In the strain-controlled test, the lower frame is slowly pushed forward at a constant rate and, if there is any strength in the specimen, the upper frame will be moved forward also. Now the upper frame rests against a proving ring; as the frame is moved forward it pushes against the proving ring. Eventually, as the proving ring becomes more compressed, the back force on the upper frame becomes so large that no further movement of the upper frame is possible. At this time the force in the proving ring is just balanced by the strength of the sample, and the peak shear strength, as a force (S_p) or as a stress (s_p) may be determined from the proving ring calibration. A load normal to the plane of shear (horizontal) is applied to the top of the specimen, allowing the applied load per unit area of specimen (σ) to be controlled very easily.

The peak shear strength of natural materials usually increases with the applied normal stress. Mohr (1900), who developed the circle of stress described earlier, suggested a general relationship between the two parameters of the form

$$s_p = f(\sigma) . \tag{3.3}$$

Often it is assumed that the relationship is linear, and long ago Coulomb (1776) proposed the equation

$$s_p = c + \sigma(\mu) = c + \sigma \tan\phi , \tag{3.4}$$

which is depicted graphically in Figure 3.7. The parameter ϕ (or μ) denotes the rate at which strength increases with the normal stress; it is usually described as the friction parameter, although there is some doubt as to whether it is really a measure of genuine inter-particle friction. The term c represents the shear strength of a specimen under zero normal stress. It is thus an indication of bonds between particles in the absence of external forces pushing the particles together. Such bonds may be due to cementing in sedimentary rocks or to physico-chemical attraction in clays. In dry, or fully saturated, sandy soil the c intercept is zero.

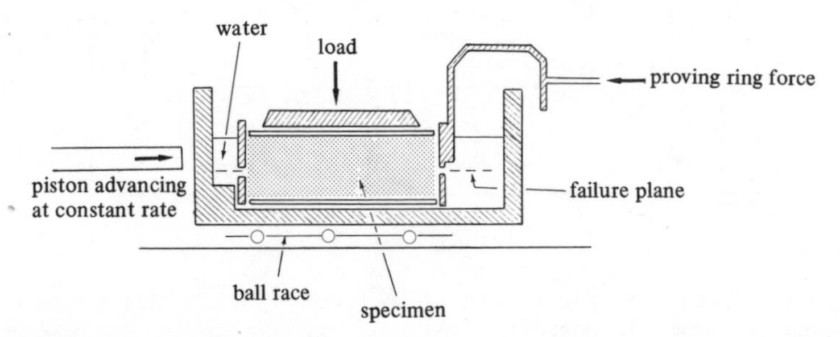

Figure 3.6. The direct-shear apparatus.

Sometimes Equation (3.4) is cited as the Mohr-Coulomb strength criterio but this is strictly incorrect as the Coulomb equation is merely a special case of the general Mohr function. It is also known as the Navier-Coulomb criterion. Evidence indicates that for small ranges of σ, the Coulomb criterion is reasonably satisfactory, but for wide ranges of σ the strength envelope (the s_p-σ plot) is distinctly curved.

For many purposes, equations describing the relationship between s_p and σ are unsatisfactory, because c and ϕ vary markedly with the conditions under which specimens are tested. Even for a material which behaves as an ideal plastic substance there is no unique value of s_p for a specified value of σ. An extra parameter, the pore pressure (the pressure of the fluid in the pore space) on the failure surface inside the specimen, at the time of failure, must be included in the relationship. This pressure (Figure 3.8) acts to dissipate the effect of the applied normal stress if it is positive (relative to atmospheric pressure) and to supplement the applied normal stress if it is negative. Equation (3.4) is therefore commonly written in the form:

$$s_p = c' + \sigma' \tan\phi' , \qquad (3.:$$

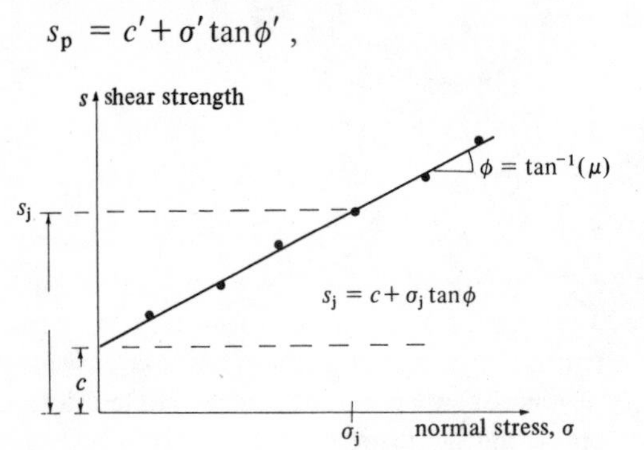

Figure 3.7. The Coulomb interpretation of shear strength.

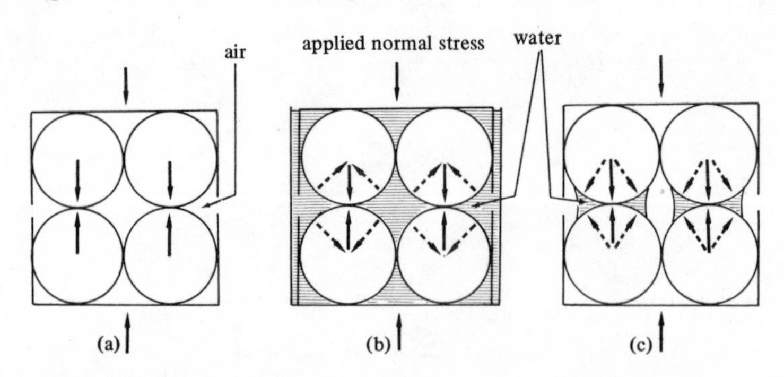

Figure 3.8. The influence of moisture on pore pressures in soil: (a) dry soil with atmospheric (zero) pore pressures; (b) saturated soil with positive pore pressures; (c) partially saturated soil with negative pore pressures.

where $\sigma' = \sigma - u$ (see Appendix 3.1 for a full development of this important point). The parameter σ' is the *effective* normal stress, σ is the applied (often called *total*) normal stress and u is the pore pressure on the failure surface at the time of failure. Equation (3.5) is a more satisfactory form of the Coulomb criterion because c' and ϕ' are much less dependent upon test conditions. As an illustration, consider a direct-shear test on a saturated coarse sand specimen which, as pointed out previously, possesses no cohesion. Figure 3.9 shows the strength envelope based on three tests at different σ values, for three types of test condition. The uppermost plot describes tests where the specimen is allowed to drain water fully whenever an increment of normal load is applied. The three points locate a straight line with ϕ equal to about 35°. In the middle plot we have assumed that, after the first increment of load has been applied, the apparatus is sealed and no drainage of water from the specimen is allowed. There is now no increase in strength as the applied normal stress is increased. The reason is that the effect of adding an increment of normal stress has been merely to raise the pore water pressure by the same amount, and the actual stress effective in mobilizing friction is unchanged.

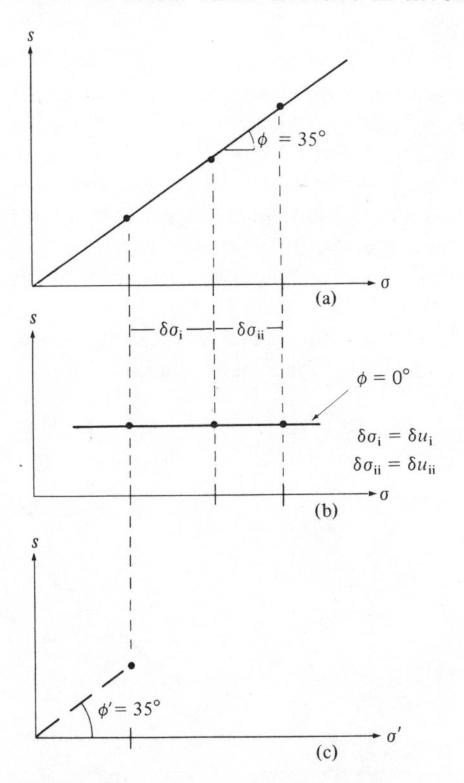

Figure 3.9. The Coulomb strength envelope in terms of total and effective normal stresses acting within a sand sample.

The results of tests with no drainage can be plotted in terms of effective stresses provided that we know the pore pressures that develop. These a shown in the lowermost plot and, because $\delta\sigma_{\mathrm{i}} = \delta u_{\mathrm{i}}$ and $\delta\sigma_{\mathrm{ii}} = \delta u_{\mathrm{ii}}$, the plots of all three tests are for the same effective normal stress value. Assuming that $c' = 0$, ϕ' as measured in the test with no drainage is the same as ϕ measured in the tests with drainage. In nature, pore pressures are positive in material below the water table and increase in magnitude with depth. (The actual relationship between pore pressure and depth depends on the pattern of groundwater flow.) If we are considering the strength at a particular point inside a hill mass, it is easily seen from Equation (3.5) that this will fluctuate with the position of the water tabl Many landslides are caused by a reduction in the shear strength along a potential failure plane by a rise in the water-table level. Note that if the water table falls below the point under study, the pores are only partiall occupied by water and the remaining pore space is filled with air. The combination of surface tension in the pore water and the curvature of th meniscus results in a decrease in pore water pressure below atmospheric pressure and, as a result, overall pore pressure is negative. This extra for pulling the particles together is termed capillary suction.

Finally we should emphasize that shear strength is also a function of t density of packing, usually expressed in terms of the *void ratio* (e), which is the volume of pore space in a specimen as a fraction of the volume o the solid particles in the specimen. Hvorslev (1960) pointed out that both cohesion and internal friction increase as the void ratio decreases and particles are more densely packed together. Many engineers, for convenien prefer to view cohesion as independent of the void ratio and treat only ϕ' as variable; this is the approach taken here. Actually, it is only peak strength and therefore peak ϕ', that is affected by the void ratio (Figure 3.10); the residual or long-term strength (and the residual ϕ' parameter) is usually independent of the initial density of packing. The reason is that, during

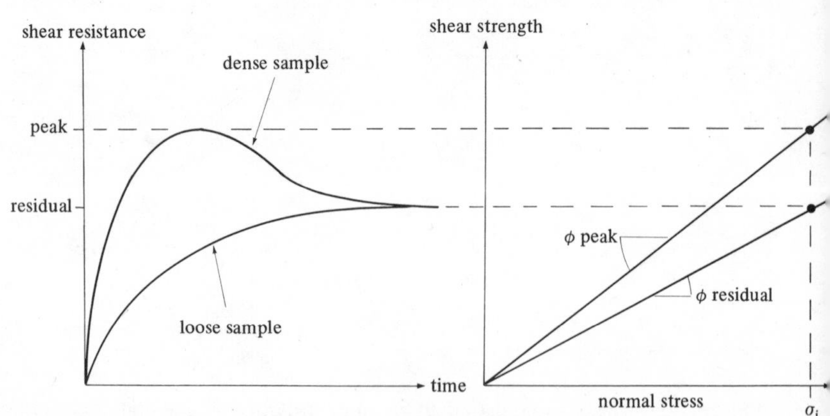

Figure 3.10. Shear test data for loose and dense sand samples tested at $\sigma = \sigma_{\mathrm{j}}$.

the process of shear, a densely-packed sample will expand and lose the extra strength due to its initially dense packing, until, ultimately, the strength falls towards that of a loosely-packed sample.

3.2 Stress-strength interrelationships

Chapter 1 and Section 3.1 provide us with the tools to examine the balance between stresses and strength along different surfaces, and at different points, within a semi-infinite material. We shall first of all consider this for a mass of cohesionless material with a flat surface, next we shall examine the slightly more complicated case of a cohesive material with a flat surface, and then we shall note the important modifications which occur with sloping surfaces.

Figure 3.11 shows the state of stress at different depths in a cohesionless material with a horizontal surface. It is evident from Figure 3.11a that the vertical stress (σ_z) at depth z is given by:

$$\sigma_z = \gamma z\,, \tag{3.6}$$

where γ is the unit weight of the material. The horizontal stress at depth z is given by

$$\sigma_x = k\gamma z\,, \tag{3.7}$$

where k is termed the *coefficient of earth pressure*. (We have previously noted that for static water $\sigma_x = \sigma_z$ so that $k = 1$.) For earth *at rest* k ranges from 0·4 to almost unity, and therefore $\sigma_z > \sigma_x$. Now σ_z and σ_x are, in Figure 3.11a, principal stresses (Chapter 1) and, because $\sigma_z > \sigma_x$, we have, for the at-rest condition, $\sigma_z = \sigma_1$ and $\sigma_x = \sigma_3$. Using this information, together with Equations (3.6) and (3.7), we can plot the Mohr circle of stress for any point, as shown in Figure 3.11b. Each circle

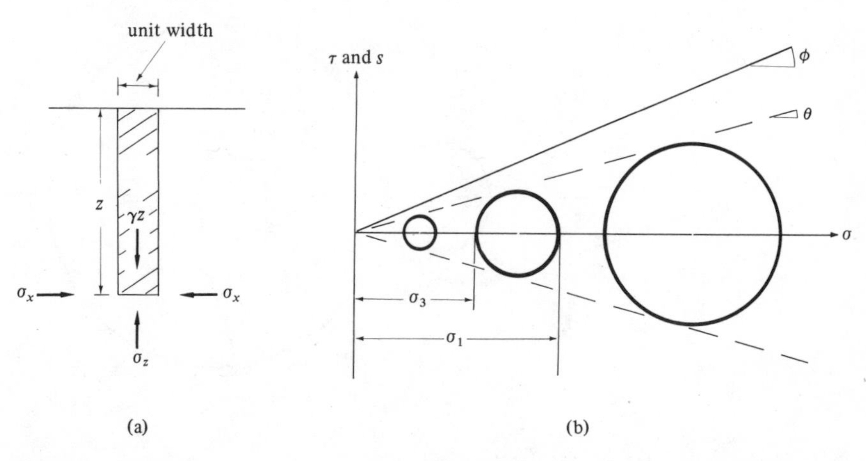

Figure 3.11. Stress conditions at a point in a cohesionless mass, with horizontal surface, at rest.

corresponds to a different depth. The radius of the circle is given by

$$r = \tfrac{1}{2}(\sigma_1 - \sigma_3) = \tfrac{1}{2}\sigma_z(1-k) \qquad (3.8)$$

so that circle radius increases linearly with depth if it is assumed that k does not vary with depth. Figure 3.11b also shows the Coulomb strength envelope for the material. In the at-rest condition the shear stress on all surfaces (different points on a circle) at all depths (different circles) is less than the maximum shear strength of the material. If we wish to bring the material to a state of incipient failure *at each point* in the mass, we must increase θ until it equals ϕ. Note that this can be undertaken in two extreme ways. One mechanism is to decrease the horizontal stress relative to σ_z (Figure 3.12); the other is to increase σ_x above σ_z so that the horizontal stress becomes the major principal stress. The former mechanism is called *active failure* and the latter termed *passive failure.* In relation to Figure 3.11a, active failure involves stretching the mass horizontally (the weight of the mass is thus *active* in this process) and passive failure involves horizontal compression. Note that, at limiting equilibrium, the coefficient k has different values depending on the mode of failure. At incipient failure, it can be shown (Appendix 3.2) that

$$k_A = \tan^2(\pi/4 - \phi/2)$$

$$k_P = \tan^2(\pi/4 + \phi/2)$$

where A and P denote the active and passive modes of failure respectively.

A more important difference between the two cases is the orientation of the shear surfaces at failure. In order to understand this we must introduce the concept of the *pole* of the Mohr circle of stress.

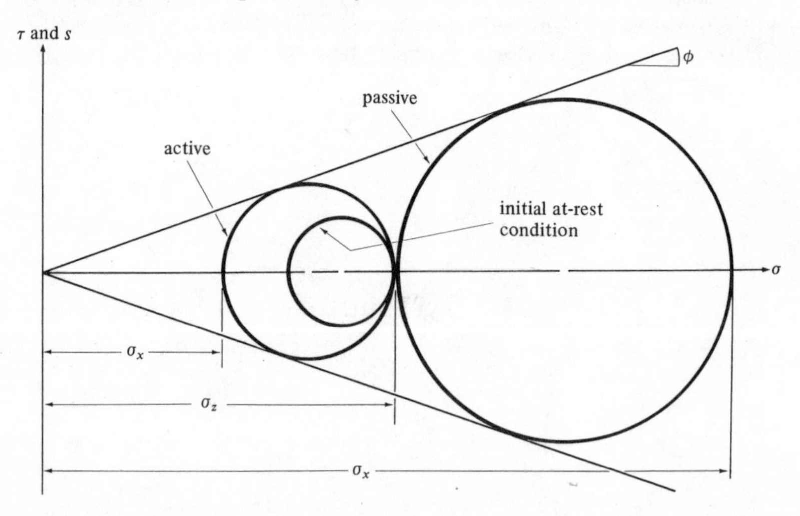

Figure 3.12. Contrast between active and passive Rankine failure conditions; σ_z held constant for simplicity.

Consider any section through a point in a soil mass inclined at an angle β to the horizontal. Assuming that the state of stress at that point on that section is known, locate the point representing that section on the Mohr circle (C) as shown in Figure 3.13. Now draw through C a line at an angle β to the horizontal. The point at which this line meets the circle is defined as the pole of the circle. It is important to bear in mind that, irrespective of the value of β, *all* lines that are drawn through a point on the Mohr circle parallel to the actual section represented by that point will pass through this pole. This may be tested by considering the major and minor principal planes. We have already noted (Chapter 1) that the section represented by C is inclined at an angle θ (as defined in the diagram) to the major principal plane. The major principal plane ($\sigma = \sigma_1$, $\tau = 0$) is represented by B in Figure 3.13. If, therefore, we construct through B a line at an angle θ to *CC′*, this should pass through the pole according to the above statement. Alternatively, the angle C′PB should equal θ. It is easy to verify from the diagram that this is, in fact, the case. Note that AP is perpendicular to BP so that it follows, further, that, in the two-dimensional case, A represents the minor principal plane ($\sigma = \sigma_3$, $\tau = 0$). Note, finally, that all that is needed to locate the pole of the Mohr circle is the state of stress on a particular plane of known inclination relative to the horizontal. Now let us return to the problem of locating the orientation of the shear planes during active and passive failure for cohesionless material with a horizontal surface.

Consider (in Figure 3.11a) the state of stress on a section parallel to the surface at some depth. The normal stress is σ_z and the shear stress is zero. At incipient active failure $\sigma_z = \sigma_1$, represented by point D on the active Mohr circle (Figure 3.14). A line through D parallel to the section represented by D (horizontal) cuts the circle again at the point (σ_3, 0); this is the active pole P_A. The passive pole P_P is determined in a similar manner. Now both circles in Figure 3.14 represent points in a mass on

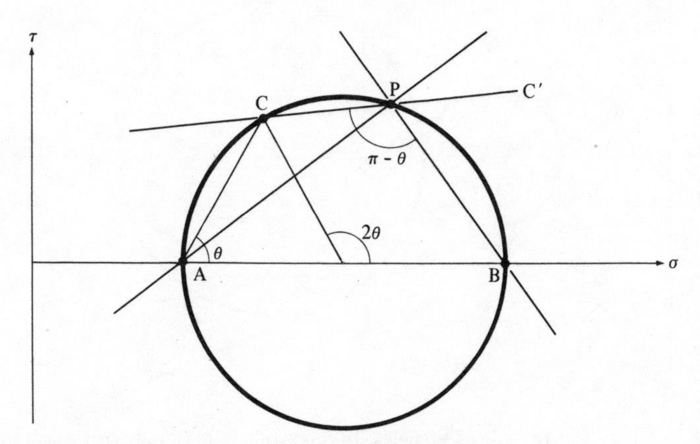

Figure 3.13. The pole of Mohr's circle of stress.

the verge of failure. The problem is to determine the orientation of the incipient failure surfaces in both cases. These failure surfaces are represented by the points F, for both active and passive failure, where the Mohr circles reach the strength envelope. At all other points on the circles (on all other sections) the shear stress is less than the maximum shear strength; at the points F, the shear stress and the shear strength are equal. Now, if we draw lines through F to the respective pole, we know (from the property of the pole mentioned above) that these lines are parallel to the actual failure surfaces in the material. From Figure 3.14, it is easily determined that in active failure, the shear planes are inclined at an angle $\pi/4+\phi/2$ with the horizontal and in passive failure this angle is $\pi/4-\phi/2$. Notice that, in both cases, the shear plane is inclined at an angle $\pi/4+\phi/2$ to the major principal plane. This will be pointed out again in Chapter 4 (particularly Appendix 4.2) in the discussion of the Culmann method of stability analysis.

The stress patterns in Figure 3.14 are usually referred to as the active and passive Rankine states of stress after the British civil engineer by that name. The active Rankine state of stress for a cohesive material with a horizontal surface is shown in Figure 3.15. The special feature here is that at depths smaller than that represented by circle C2 the horizontal stress in the active state is negative, indicating tension in the horizontal direction. It can be seen from the geometry of circle C2 (Appendix 4.5) that tension disappears at a depth given by

$$Z_0 = \frac{2c}{\gamma}\tan(\pi/4+\phi/2)\,. \tag{3.9}$$

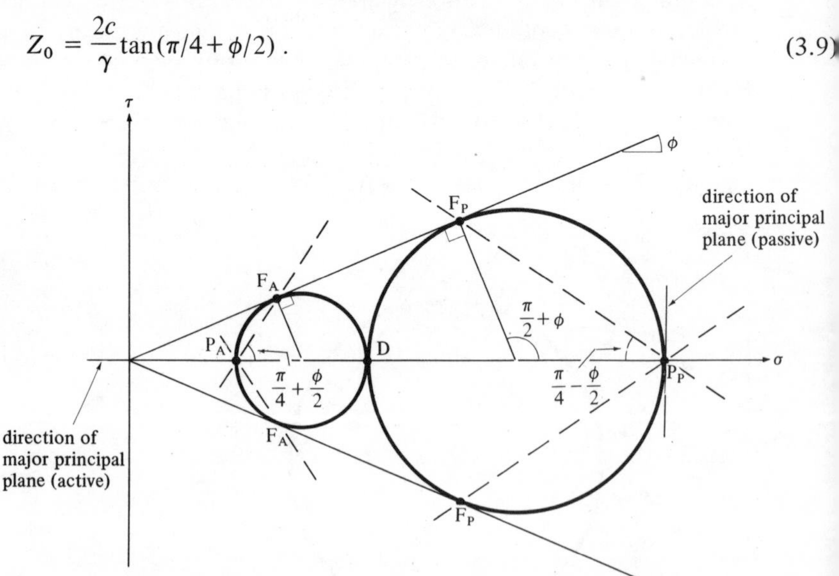

Figure 3.14. Orientation of failure surfaces in cohesionless material with a horizontal surface; dashed lines indicate directions of failure surfaces.

Note that the orientation of the shear planes in cohesive material with a horizontal surface is, in terms of ϕ, identical to the previous case.

Now consider the failure surfaces in the Rankine states of stress for cohesionless material with a sloping surface (Figure 3.16). The stresses on a section parallel to the surface are now $\sigma = \sigma_z \cos^2 i$ and $\tau = \sigma_z \cos i \sin i$,

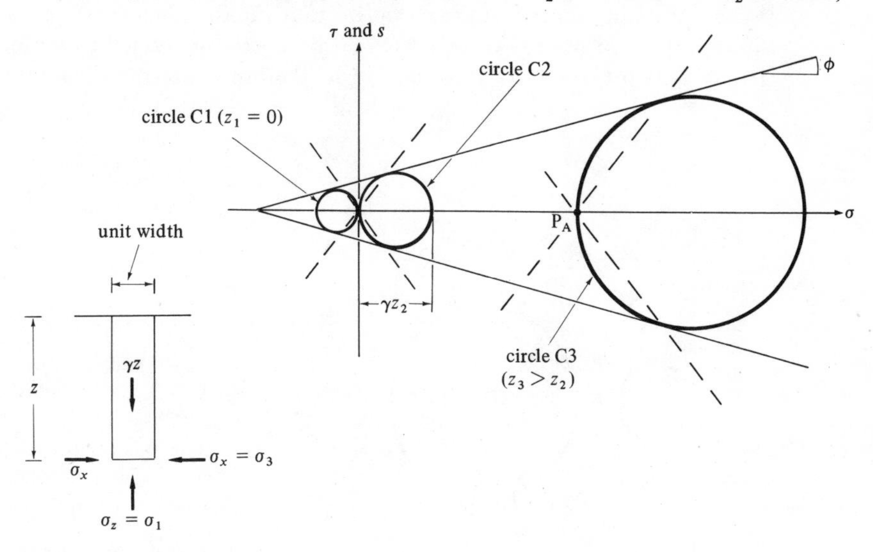

Figure 3.15. Conditions at active failure in cohesive material with a horizontal surface.

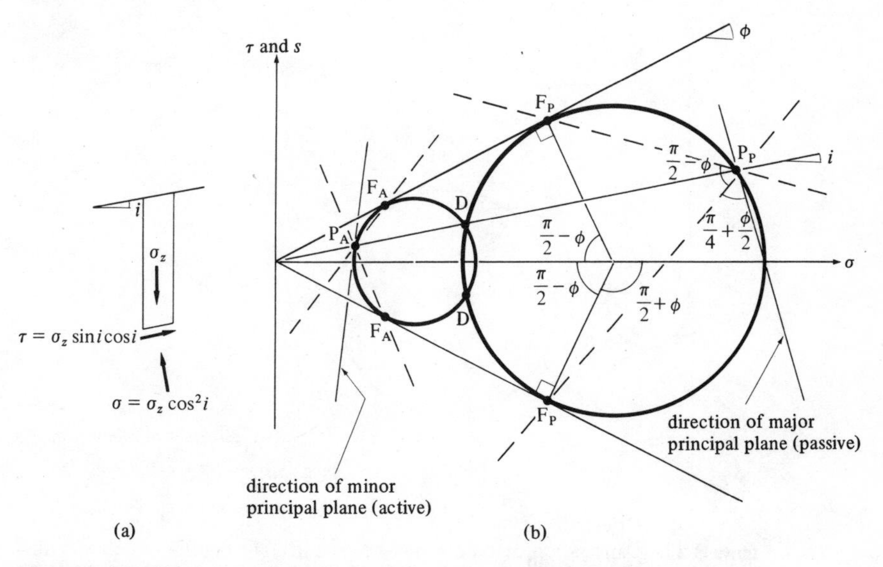

Figure 3.16. Stress conditions at active and passive failure in cohesionless material with a sloping surface.

where i is the angle of the surface slope. Note that the ratio τ/σ is in this case equal to $\tan i$. Thus the point D on the Mohr circle representing the state of stress, at any depth, on a section parallel to that surface is located on a line inclined at an angle i with the σ axis, as indicated in the diagram. Now let us locate the active and passive poles; if we draw a line through D parallel to the section represented by that point, this line will pass through the two poles. Clearly such a line is coincident with the line already drawn through D at an angle i to the horizontal. The poles are

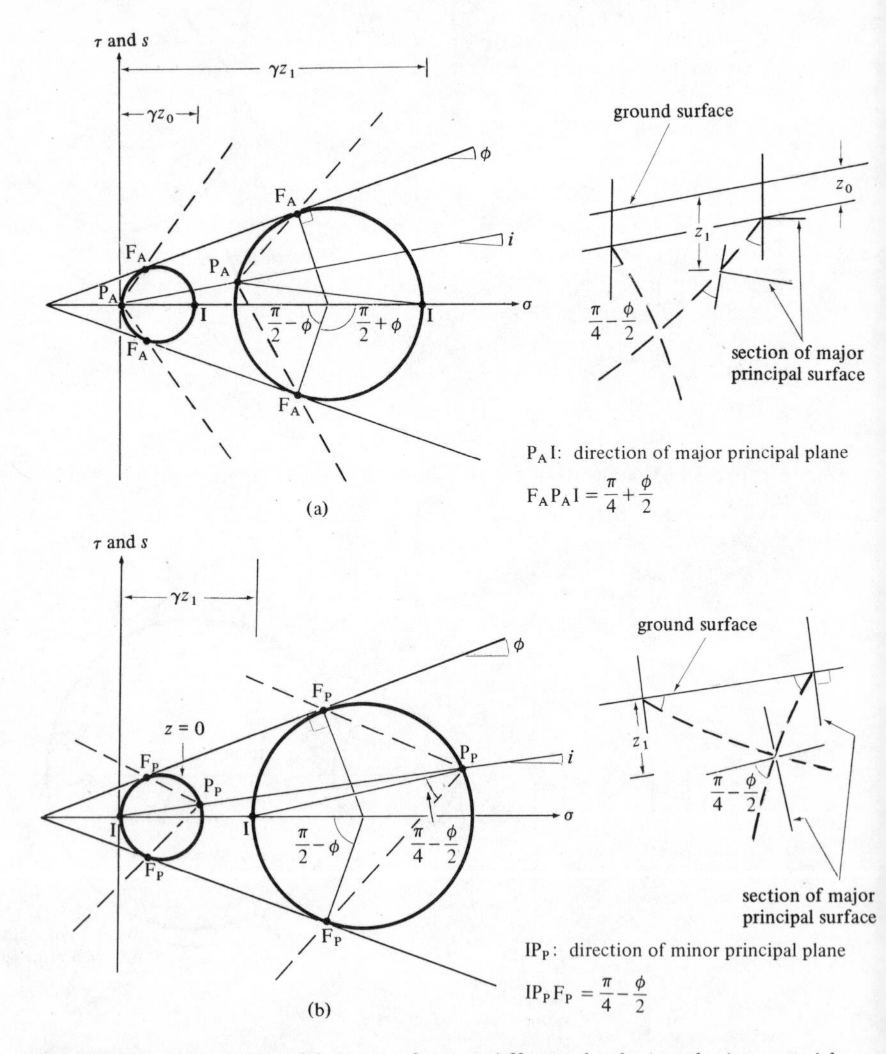

Figure 3.17. Orientation of failure surfaces at different depths in cohesive material with a sloping surface ($i < \phi$): (a) during active failure; (b) during passive failure; as $z \to \infty$, shear surfaces approach pattern for cohesionless material.

thus no longer on the σ axis as they were for masses with a horizontal surface. This produces a different orientation of the shear planes relative to the horizontal. Note that, because the major principal plane is no longer horizontal (active) or vertical (passive), the shear planes still make an angle $\alpha = \pi/4 + \phi/2$ with it (Figure 3.16b).

The Rankine states of stress for a cohesive mass with a sloping surface are shown in Figure 3.17 for the case where $\phi > 0$ and $i < \phi$. Notice that, because the strength envelope and the pole-line do not intersect the σ axis at the same point, the directions of the failure surfaces are now affected by the position of the Mohr circle. This means that, unlike the three previous cases, the shear surfaces change orientation with depth. Notice the difference between the failure surfaces in the passive and active conditions; this will be referred to in Chapter 5 in the discussion of glacier flow patterns. Finally, let us emphasize that in all four cases discussed so far the *entire mass* has been brought to the state of incipient failure. For slopes where $i > \phi$ similar conditions may prevail, provided that the mass is thinner than a critical thickness. As soon as this thickness is attained, failure along a single distinct surface will occur, irrespective of the state of stress (active, passive or at-rest) in the material. In a semi-infinite mass (Figure 3.18) this will take the form of a slide along a plane parallel to the surface. Similar instability for finite slopes is discussed in Chapter 4.

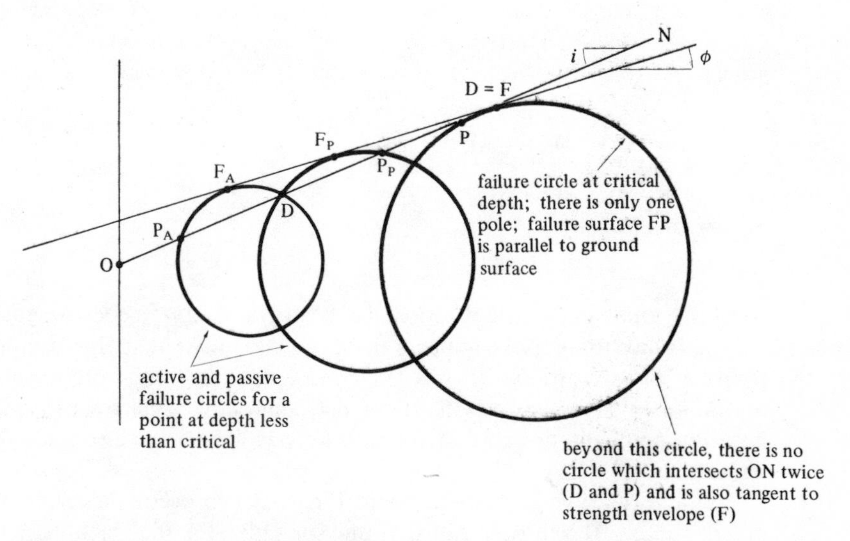

Figure 3.18. The critical depth for stability in a semi-infinite cohesive mass of steepness greater than the ϕ parameter.

3.3. Stress-strain-time relationships

Some mention has already been made of the nature of stress-strain rate relationships in the discussion of strength in Section 3.1. Because this topic is of fundamental importance in understanding the failure characteristics, particularly the velocity profiles with depth, of a mass-movement of water, soil, ice or some other material, we shall pursue it now in more detail. Previously we discussed strain and strain rate solely in the context of the compression [triaxial ($\sigma_3 > 0$) or uniaxial ($\sigma_3 = 0$)] of a soft rock such as clay; the term *linear strain* is given to this type of deformation. An alternative form of strain, involving angular distortion rather than linear displacement, is shown in Figure 3.19b. This is termed *shear strain*. The reader is warned that in the existing solid mechanics literature various notations and measures of shear strain are employed. Early workers, e.g. Timoshenko and Goodier (1951), Hoffman and Sachs (1953), defined shear strain in terms of the *full* angular deformation of the body that is strained, that is:

$$\gamma_{xz} = \gamma_{zx} = \frac{\partial u_x}{\partial z} + \frac{\partial u_z}{\partial x}$$

where u_x and u_z are orthogonal components of the displacement vector $\mathbf{u}$ (see Figure 3.19b), or, for small angles

$$\gamma_{xz} = \gamma_{zx} = \theta$$

with θ in radians. More recently others, e.g. Mendelson (1968) whom we follow here, have preferred for convenience in certain mathematical procedures to define shear strain in terms of half the angular change, that is

$$\epsilon_{xz} = \epsilon_{zx} = \frac{1}{2}\left(\frac{\partial u_x}{\partial z} + \frac{\partial u_z}{\partial x}\right)$$

or, for small angles,

$$\epsilon_{xz} = \epsilon_{zx} = \theta/2\,.$$

Although the change in definition has been frequently accompanied by a change in notation, there appears to be no statement that this should be regarded as convention. In this text, while adopting the more recent definition of shear strain in terms of half angles, we shall retain the old notation to facilitate ready distinction between shear (γ) and normal (ϵ) strains.

The deformation process shown in Figure 3.19b occurs in solids, but not in fluids. It will be recalled from Appendix 1.1 that orthogonal shear stresses in solids are equal, e.g. $\tau_{zx} = \tau_{xz}$. The same is true for orthogonal shear strains, e.g. $\gamma_{zx} = \gamma_{xz}$. In the case of moving fluids, however, this situation cannot exist because the only non-zero shear stress, and thus the only non-zero shear strain, occurs in the direction of flow.

Such a situation is shown in Figure 3.19c. The non-zero shear stress is τ_{zx} and the non-zero shear strain is $\gamma_{zx}(=\theta)$. The process of deformation in Figure 3.19c is called *simple shear*. It may be visualized by analogy with the sliding of cards over each other in a pack of playing cards. Geometrically, it involves a translation, about a constant origin, from old two-dimensional coordinates (x, z) to new ones (x^*, z^*) where $x^* = x + z\tan 2\theta$ and $z^* = z$. (Simple shear should not be confused with

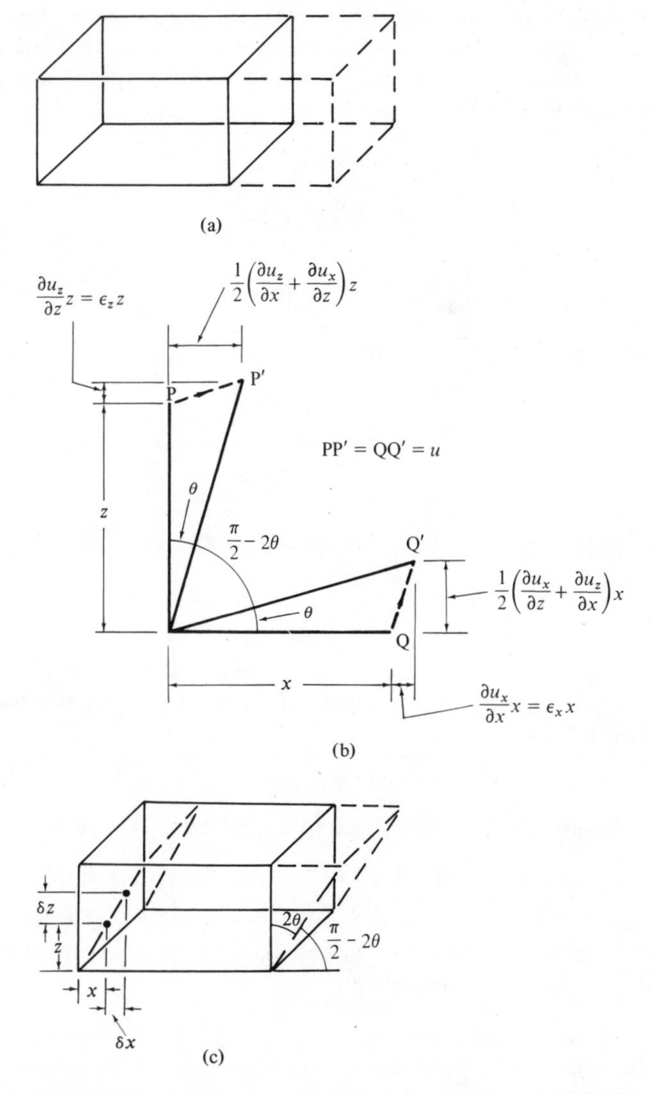

Figure 3.19. Alternative modes of strain: (a) linear strain; (b) shear strain; (c) simple shear strain.

pure shear which involves a combination of elongation and shortening, in mutually orthogonal directions, such that the volume of the body is unchanged. Geometrically, pure shear involves a translation to new coordinates given by $x^* = kx$ and $z^* = k^{-1}z$.)

Although both simple shear and uniaxial compression are easily simulated in the laboratory, many field problems involve complex deformations combining both shear and linear strain. We therefore need a general relationship between stress and strain (or strain rate) which can readily be reduced to special cases such as uniaxial tension. In Chapter 1 we introduced the concept of the octahedral plane, and the stress-strain relationships on this plane are commonly used for this purpose. Let us recall Equations (1.9)-(1.11)

$$\sigma_{oct} = \tfrac{1}{3}(\sigma_1 + \sigma_2 + \sigma_3) = \sigma_m$$

$$\tau_{oct} = \frac{1}{\sqrt{3}}[(\sigma_1 - \sigma_m)^2 + (\sigma_2 - \sigma_m)^2 + (\sigma_3 - \sigma_m)^2]^{1/2}$$

or

$$\tau_{oct} = \tfrac{1}{3}[(\sigma_1 - \sigma_2)^2 + (\sigma_2 - \sigma_3)^2 + (\sigma_3 - \sigma_1)^2]^{1/2} .$$

Note that for uniaxial compression or tension ($\sigma_1 > 0$, $\sigma_2 = \sigma_3 = 0$) Equations (1.9) and (1.11) reduce to

$$\sigma_{oct} = \tfrac{1}{3}\sigma_1 \tag{3.10}$$

$$\tau_{oct} = \frac{\sqrt{2}}{3}\sigma_1 \tag{3.11}$$

and, for simple shear (τ_{zx} being the only non-zero stress), we obtain (Appendix 3.3)

$$\tau_{oct} = \sqrt{\tfrac{2}{3}}\,\tau_{zx} . \tag{3.12}$$

Similarly it can be shown (Appendix 3.4) that the *octahedral shear strain* (γ_{oct}) can be expressed in terms of the principal strains (linear strain along the principal strain axes) as follows

$$\gamma_{oct} = \tfrac{1}{3}[(\epsilon_1 - \epsilon_2)^2 + (\epsilon_2 - \epsilon_3)^2 + (\epsilon_3 - \epsilon_1)^2]^{1/2} \tag{3.13}$$

and that the octahedral shear strain rate ($d\gamma_{oct}/dt = \dot{\gamma}_{oct}$) is given by

$$\dot{\gamma}_{oct} = \tfrac{1}{3}[(\dot{\epsilon}_1 - \dot{\epsilon}_2)^2 + (\dot{\epsilon}_2 - \dot{\epsilon}_3)^2 + (\dot{\epsilon}_3 - \dot{\epsilon}_1)^2]^{1/2} . \tag{3.14}$$

These equations will be encountered again in the discussion of glacier movement in Chapter 5. Note that, in uniaxial tension or compression, Equations (3.13) and (3.14) reduce to

$$\gamma_{oct} = \frac{1}{\sqrt{2}}\epsilon_1 \tag{3.15}$$

$$\dot{\gamma}_{oct} = \frac{1}{\sqrt{2}}\dot{\epsilon}_1 \tag{3.16}$$

and, for the simple shear case, the equations reduce to

$$\gamma_{oct} = \sqrt{\tfrac{2}{3}}\gamma_{zx} \tag{3.17}$$

$$\dot{\gamma}_{oct} = \sqrt{\tfrac{2}{3}}\dot{\gamma}_{zx}\ ; \tag{3.18}$$

these equations are derived in Appendix 3.5.

So far it may not be clear why the stress-strain rate behaviour is so important; the reason is that it is closely related to the velocity profiles of moving substances and from Chapter 2 we have already seen that the velocity profile of moving fluids is very important in determining erosive ability. Suppose that we subject a small cube of viscous material to the process of simple shear by the application of a shear stress τ_{zx}. The form of the cube after a small period of time is shown in Figure 3.19c. The amount of shear strain is indicated by $\tan\theta$ or, for small angles, by θ (in radians), and the rate of shear strain is given by

$$\dot{\gamma}_{zx} = \frac{\mathrm{d}\theta}{\mathrm{d}t}\ . \tag{3.19}$$

Now, from the diagram, $2\theta = \tan 2\theta = \delta x/\delta z$ for small angles; substituting for θ in Equation (3.19) we have

$$\dot{\gamma}_{zx} = \frac{\mathrm{d}}{\mathrm{d}t}\left(\frac{1}{2}\frac{\mathrm{d}x}{\mathrm{d}z}\right) = \frac{1}{2}\frac{\mathrm{d}}{\mathrm{d}z}\left(\frac{\mathrm{d}x}{\mathrm{d}t}\right).$$

But $\mathrm{d}x/\mathrm{d}t$ is the velocity of a point at a height $(z+\delta z)$ above the basal plane relative to the velocity of a point at a height z above the basal plane. If we denote this velocity difference over the distance δz by δu, we have

$$\dot{\gamma}_{zx} = \frac{1}{2}\frac{\mathrm{d}}{\mathrm{d}z}(\delta u) = \frac{1}{2}\frac{\mathrm{d}u}{\mathrm{d}z}\ . \tag{3.20}$$

The term $\dot{\gamma}_{zx}$ is thus a measure of the velocity gradient (with depth) of the material under deformation. In this example, the velocity gradient and the shear strain rate are constant with depth; in natural materials this is unlikely to be the case. If, however, we can calculate the relationship between shear strain rate and depth, we can compute the velocity profile of the movement. In order to do this we need to make the (plausible) assumption that the shear stress is a function of depth and, from the stress-strain rate curve, obtain the strain rate-depth relationship. The relationship between shear stress and depth for an infinite slope of some material is shown in Figure 3.20 (note that the axes have been rotated from the position of the previous diagram). From the figure,

$$\tau_{zx} = \rho g(D-z)\sin i\ . \tag{3.21}$$

Now, if the relationship between strain rate and stress is expressed in the form

$$\dot{\gamma}_{zx} = \overline{A}\tau_{zx}^{n} \tag{3.22}$$

it is possible, by substituting for τ_{zx} from Equation (3.21), to develop an equation linking the velocity gradient with depth.

From Equations (3.20), (3.21), and (3.22), we have

$$\frac{1}{2}\frac{du}{dz} = \overline{A}(\rho g)^n (D-z)^n \sin^n i$$

or

$$\frac{du}{dz} = B(D-z)^n ,$$

where $B = 2\overline{A}(\rho g)^n \sin^n i$. Integration yields

$$u = -\frac{B}{n+1}(D-z)^{n+1} + C , \qquad (3.23)$$

provided that B is independent of z. When $z = 0$, $u = u_b$ (basal velocity) and Equation (3.23) becomes

$$u_b = -\frac{B}{n+1}D^{n+1} + C , \qquad (3.24)$$

and when $z = D$, $u = u_s$ (surface velocity) we have

$$u_s = C . \qquad (3.25)$$

Let us now consider the implications of these equations for various materials described in Figure 3.1.

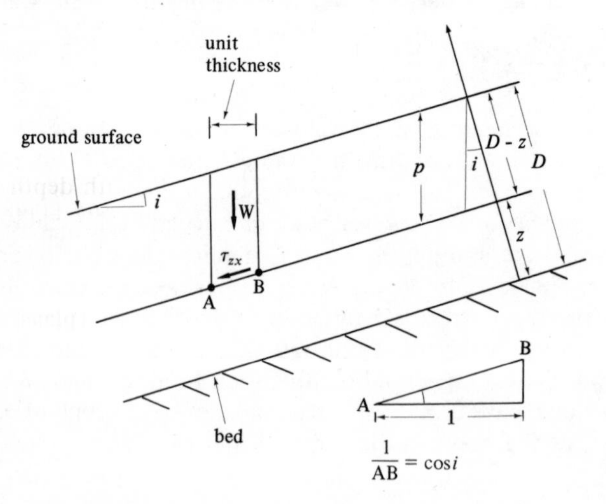

$$\tau_{zx} = \frac{W \sin i}{AB} = \rho g p \sin i \cos i = (\rho g)(D-z)\sin i$$

Figure 3.20. The increase in the shear stress with depth on an infinite slope.

(a) *Ideal Newtonian fluid*
From Figure 3.1, we know that Equation (3.22) is linear, and it may be written

$$\dot{\gamma}_{zx} = \frac{1}{\mu}\tau_{zx}\,, \tag{3.22a}$$

where μ is the dynamic molecular viscosity of laminar flow. If we substitute $\bar{A} = 1/\mu$ and $n = 1$ in Equation (3.23), this yields

$$u = -\tfrac{1}{2}B(D-z)^2 + C\,.$$

Now, from Chapter 2, we know that $u_b = 0$, so that, from Equation (3.24),

$$C = \tfrac{1}{2}BD^2$$

and therefore,

$$u = \tfrac{1}{2}B[D^2 - (D-z)^2]$$

or

$$u = -\tfrac{1}{2}B(2D-z)z \tag{3.26}$$

which is, of course, the parabolic velocity profile (Figure 3.21) for laminar flow noted in Chapter 2. In turbulent flow conditions μ must be replaced by η (the dynamic eddy viscosity coefficient) which is a function of velocity and, therefore, a function of depth also. Accordingly, the development of a velocity profile for turbulent flow is more complex, as noted previously.

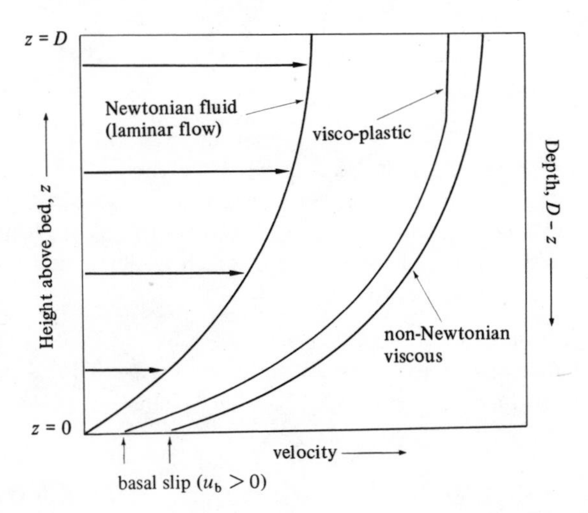

Figure 3.21. Velocity profiles for various materials on an infinite slope.

(b) *Non-Newtonian viscous material*
Although glaciologists describe ice as pseudo-plastic material, for many purposes it can be treated in the present category. Glen's (1952) data for polycrystalline ice at −0·8°C to −2·5°C (Figure 3.5), under uniaxial compression, conformed to an equation of the form

$$\dot{\epsilon}_1 = A'\sigma_1^n$$

with $A' = 0 \cdot 0074$ and $n = 4$ (the value of n is unaffected by the type of flow law) when $\dot{\epsilon}_1$ is in units of percent per year and σ_1 is expressed in bars. (As we shall note in Chapter 5, the flow law for glacial ice has been modified appreciably since Glen's 1952 paper, but these data are adequate for present purposes.) Substitution for n in Equation (3.23) yields

$$u = -\tfrac{1}{5}B(D-z)^4 + C\,, \qquad (3.27)$$

bearing in mind that A' (for uniaxial compression) must be converted to $\overline{A}$ (for simple shear), through the use of Equations (3.11), (3.12), (3.16), and (3.18), before B can be determined. Note that, in the case of ice, we cannot assume $u_b = 0$; this is, therefore, the major difference between the velocity profiles of ice and water (Figure 3.21), together with the higher value of n for ice.

(c) *Visco-plastic material*
The existence of a yield stress in this type of material means that the velocity gradient is zero in the uppermost parts of the flow (Figure 3.21). The reason is as follows. Shear strain, and therefore shear strain rate, is zero at shear stresses less than the yield stress τ_y. From Equation (3.21), the shear stress is a function of depth below the surface $(D-z)$; thus for values of $(D-z)$ given by

$$D - z < \frac{\tau_y}{\rho g \sin i} \qquad (3.28)$$

the velocity gradient will be zero. Often glacial ice is regarded as possessing a yield strength of about 1 kg cm^{-2}; if ρg for ice is taken as ~1 g cm^{-3}, one would expect from (3.28) a zero velocity gradient down to depths of about $10/\sin i$ metres. Evidence (see Chapter 5) suggests this is not the case, supporting the laboratory data which indicate that the yield strength of ice is virtually zero.

(d) *Ideal rigid plastic material*
The stress-strain behaviour of this type of material (Figure 3.1) can also be expressed in the form of Equation (3.22). This yields

$$\dot{\gamma}_{zx} = \tau_y^{-\infty}\tau_{zx}^{\infty} \qquad (3.22b)$$

where τ_y is the yield stress in shear. For $\tau_{zx} < \tau_y$, $\dot{\gamma}_{zx} \rightarrow 0$ and for $\tau_{zx} > \tau_y$, $\dot{\gamma}_{zx} \rightarrow \infty$. The velocity profile describing this type of behaviour corresponds to that of a slide rather than a flow (Figure 3.22). This is a common mode of failure of many soil mantles on hillside slopes: the mantle is essentially immobile ($\tau_{zx} < \tau_y$ at all depths) until, for some reason, the shear strength decreases so much that the yield strength becomes less than the shear stress on the basal plane (Figure 3.22), at which time a landslide occurs. Not all soils behave in this manner, however, as pointed out in Section 3.1.

With this elementary background on the mechanics of solid materials, we are now in a position to examine the mechanics of erosion associated with the mass movement of rock, soil, and glacial ice. Two points should be emphasized, however, before leaving the topic of stress-strain rate relationships. Flow laws, such as Equation (3.22), are not invariant with temperature. Under colder conditions, viscosity may be expected to increase, affecting the n exponent and, particularly, the coefficient $\overline{A}$. In the above derivation of velocity profiles from the flow laws it has been implicitly assumed that temperature is constant with depth. For thick masses this assumption may have to be abandoned, as we shall note in Chapter 5 for polar ice masses. Another assumption made above is that the flow law is unaffected by the level of the hydrostatic stress (the average normal stress at a point) which, of course, increases with depth. Although this is a common assumption in the literature, it has not been unequivocally demonstrated.

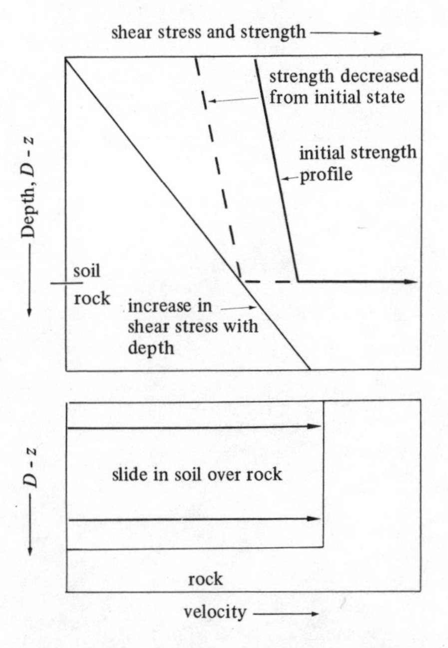

Figure 3.22. Failure of ideal rigid plastic material on an infinite slope.

Bibliography

Bishop, A. W., Eldin, G., 1950, "Undrained triaxial tests on saturated sands and their significance in the general theory of shear strength", *Géotechnique,* **2**, 13-32.

Coulomb, C. A., 1776, "Essai sur une application des règles des maximis et minimis à quelques problèmes de statique relatifs à l'architecture", *Mémoirs présentées par divers savants* (Academie des Sciences, Paris).

Glen, J. W., 1952, "Experiments on the deformation of ice", *J. Glaciol.,* **2**, 111-114.

Hoffman, O., Sachs, G., 1953, *Introduction to the Theory of Plasticity for Engineers* (McGraw-Hill, New York).

Hvorslev, M. J., 1960, "Physical components of the shear strength of saturated clay", *Proc. Am. Soc. Civil Engrs. Conference on Shear Strength of Cohesive Soils.*

Mendelson, A., 1968, *Plasticity: Theory and Application* (MacMillan, New York).

Mohr, O., 1900, "Die Elastizitätsgrenze und Bruch eines Materials", *Z. Ver. dt. Ing.,* **44**, 1524.

Taylor, D. W., 1948, *Fundamentals of Soil Mechanics* (John Wiley, New York).

Terzaghi, K., 1943, *Theoretical Soil Mechanics* (John Wiley, New York).

Timoshenko, S., Goodier, J. N., 1951, *Theory of Elasticity* (McGraw-Hill, New York).

Wu, T. H., 1966, *Soil Mechanics* (Allyn and Bacon, Boston).

Yong, R. N., Warkentin, B. P., 1966, *Introduction to Soil Behaviour* (MacMillan, New York).

Appendix 3

3.1 Development of the principle of effective stress

Figure A3.1 attempts to show the nature of intergranular contacts, in plan and in section, along a failure surface. Let the area of this surface be denoted by A, and the much smaller area of contact between particles on this plane by A_s; let the fraction of the area occupied by intergranular contacts be a $(= A_s/A)$. If the intergranular stress is denoted by σ_i, the intergranular force on the plane is equal to $\sigma_i A_s$. The effective normal stress σ' is defined, according to soil mechanics practice, as the *average* intergranular force *per unit area* of the plane: $\sigma' = \sigma_i A_s/A$. Resolving forces vertically, at equilibrium, we obtain

$$\sigma A = \sigma_i A_s + u(A - A_s)$$

$$\sigma = \sigma_i \frac{A_s}{A} + u\left(1 - \frac{A_s}{A}\right)$$

or

$$\sigma' = \sigma - u(1-a) . \tag{A3.1}$$

The value of a is not accurately known, but experiments by Bishop and Eldin (1950) have indicated that, for sands, it is less than a few percent. Accordingly, for sands, Equation (A3.1) may be approximated by

$$\sigma' = \sigma - u \tag{A3.2}$$

without serious loss of accuracy. The situation is probably much more complicated for clay soils. There is much speculation that solid-to-solid contacts do not exist in clays. Moreover, there is also doubt as to whether, under field conditions, it is possible to saturate a clay soil completely.

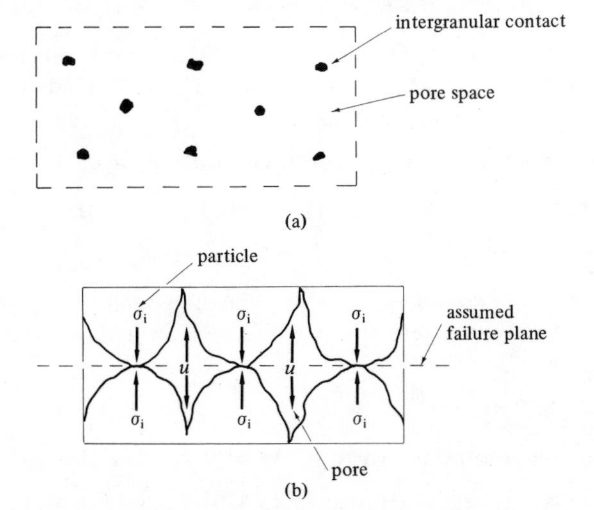

Figure A3.1. The nature of interparticle contact on a failure plane: (a) plan view of part of failure plane; (b) section through part of failure zone.

3.2 Determination of values of the coefficient of earth pressure at active and passive Rankine failure

Referring to Figure 3.11a, at failure

$$\sin\phi = \sin\theta = \frac{\frac{1}{2}(\sigma_1 - \sigma_3)}{\sigma_3 + \frac{1}{2}(\sigma_1 - \sigma_3)} .$$

Substituting $\sigma_1 = \sigma_z$, $\sigma_3 = \sigma_x$, for the *active* case, we obtain

$$\sin\phi = \frac{\frac{1}{2}(\sigma_z - \sigma_x)}{\sigma_x + \frac{1}{2}(\sigma_z - \sigma_x)}$$

or, with $\sigma_x = k\sigma_z$,

$$\sin\phi = \frac{1-k}{1+k} ,$$

which, on rearranging, yields

$$k_{\mathrm{A}} = \frac{1-\sin\phi}{1+\sin\phi} = \tan^2(\pi/4 - \phi/2) . \qquad \text{(A3.3)}$$

Similarly, substitution for the passive case, yields

$$k_{\mathrm{P}} = \tan^2(\pi/4 + \phi/2) . \qquad \text{(A3.4)}$$

From (A3.3) and (A3.4), we obtain:

$$0 < k_{\mathrm{A}} \leqslant 1 \leqslant k_{\mathrm{P}} .$$

3.3 Shear stress on the octahedral plane in terms of general coordinates (*x, y, z*)

The octahedral shear stress is given, in terms of general coordinates, by

$$\tau_{\mathrm{oct}} = \tfrac{1}{3}[(\sigma_x - \sigma_y)^2 + (\sigma_y - \sigma_z)^2 + (\sigma_z - \sigma_x)^2 + 6(\tau_{xy}^2 + \tau_{yz}^2 + \tau_{zx}^2)]^{\frac{1}{2}} \qquad \text{(A3.5)}$$

which readily reduces to Equation (1.11) when translated into principal coordinates. The derivation of (A3.5) may be found in any standard text on plasticity. An alternative form of (A3.5) is

$$\tau_{\mathrm{oct}}^2 = \tfrac{2}{9}[(\sigma_x^2 + \sigma_y^2 + \sigma_z^2) + 3(\tau_{xy}^2 + \tau_{yz}^2 + \tau_{zx}^2) - (\sigma_x\sigma_y + \sigma_y\sigma_z + \sigma_z\sigma_x)]$$

although note that Mendelson (1968, p.37) *mistakenly* writes this as

$$\tau_{\mathrm{oct}}^2 = \tfrac{2}{9}[(\sigma_x + \sigma_y + \sigma_z)^2 + 3(\tau_{xy}^2 + \tau_{yz}^2 + \tau_{zx}^2) - (\sigma_x\sigma_y + \sigma_y\sigma_z + \sigma_z\sigma_x)] .$$

Equation (A3.5) simplifies considerably in uniaxial stress and simple shear. In uniaxial stress, the only non-zero stress in (A3.5) is σ_z and, accordingly,

$$\tau_{\mathrm{oct}} = \frac{\sqrt{2}}{3}\sigma_z . \qquad \text{[Equation (3.11)]}$$

In simple shear, the only non-zero stress in (A3.5) is τ_{zx}, and this yields

$$\tau_{\mathrm{oct}} = \tfrac{1}{3}\sqrt{6}\tau_{zx} = \sqrt{\tfrac{2}{3}}\tau_{zx} . \qquad \text{[Equation (3.12)]}$$

3.4 Shear strain on the octahedral plane

Strain may be described in two general ways known as the *Lagrangian* and the *Eulerian* approaches. The Lagrangian method defines the amount of deformation relative to the *initial* geometry of a body before strain has occurred. The Eulerian approach describes the deformation in terms of the *final* geometry after strain has taken place. The distinction is shown in Figure A3.2. For infinitesimal strains the two approaches give identical results. At large strains (deferred until Appendix 5.1) the two approaches give different answers. In discussing the octahedral strain components, we shall use the Eulerian approach (Figure A3.2b) following Mendelson (1968, p.56).

The strain in the diagram consists of two parts: a linear strain ϵ (= RQ/A) and a shear strain θ (= RP/A). The linear strain is determined as follows. From elementary geometry,

$$(A+\mathrm{RQ})^2 = (A_1+\mathrm{d}A_1)^2+(A_2+\mathrm{d}A_2)^2+(A_3+\mathrm{d}A_3)^2 \; ;$$

for infinitesimal strains, the squares of small quantities (such as RQ, dA, etc.) may be neglected and, on expansion, we obtain

$$A^2+2A\,\mathrm{RQ} \approx A_1^2+A_2^2+A_3^2+2(A_1\mathrm{d}A_1+A_2\mathrm{d}A_2+A_3\mathrm{d}A_3)$$

or, noting that $A^2 = A_1^2+A_2^2+A_3^2$, we obtain

$$\frac{\mathrm{RQ}}{A} \approx \frac{A_1\mathrm{d}A_1}{A^2}+\frac{A_2\mathrm{d}A_2}{A^2}+\frac{A_3\mathrm{d}A_3}{A^2} \; .$$

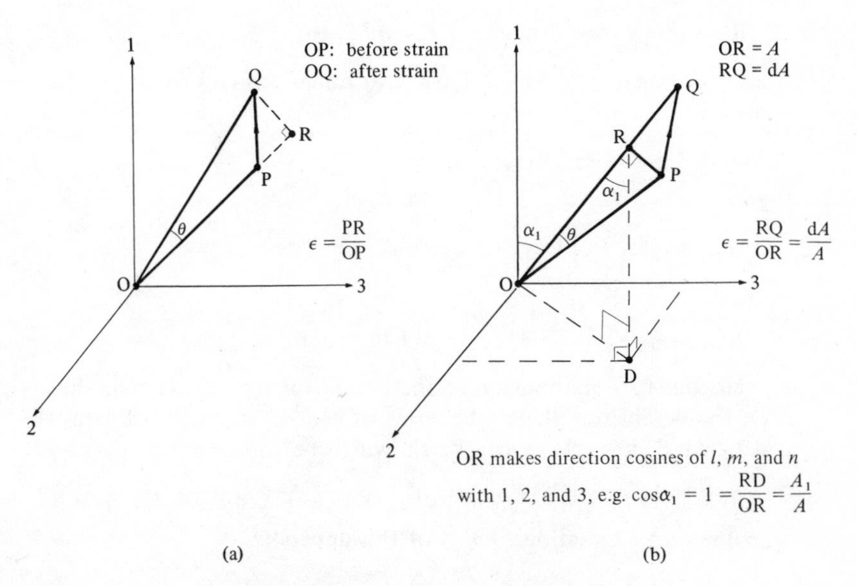

Figure A3.2. Alternative definitions of strain: (a) Lagrangian; (b) Eulerian.

This equation involves the principal strains ϵ_1 $(= \mathrm{d}A_1/A_1)$, ϵ_2 $(= \mathrm{d}A_2/A_2)$, and ϵ_3 $(= \mathrm{d}A_3/A_3)$; it may thus be rewritten

$$\frac{\mathrm{RQ}}{A} = \epsilon = \left(\frac{A_1}{A}\right)^2 \epsilon_1 + \left(\frac{A_2}{A}\right)^2 \epsilon_2 + \left(\frac{A_3}{A}\right)^2 \epsilon_3$$

or, in terms of the direction cosines (from the diagram),

$$\epsilon = l^2\epsilon_1 + m^2\epsilon_2 + n^2\epsilon_3 \,. \tag{A3.6}$$

In order to determine the shear strain θ, refer to the geometry of ΔRPQ and note that

$$\mathrm{RQ}^2 + \mathrm{RP}^2 = \mathrm{PQ}^2$$

or

$$\frac{\mathrm{RQ}^2}{A^2} + \frac{\mathrm{RP}^2}{A^2} = \frac{\mathrm{PQ}^2}{A^2}$$

or

$$\epsilon^2 + \theta^2 = \frac{\mathrm{PQ}^2}{A^2} = \left(\frac{\mathrm{d}A_1}{A}\right)^2 + \left(\frac{\mathrm{d}A_2}{A}\right)^2 + \left(\frac{\mathrm{d}A_3}{A}\right)^2$$

whence, using $\mathrm{d}A_1 = \epsilon_1 A_1$ etc.,

$$\epsilon^2 + \theta^2 = l^2\epsilon_1^2 + m^2\epsilon_2^2 + n^2\epsilon_3^2$$

or

$$\theta^2 = l^2\epsilon_1^2 + m^2\epsilon_2^2 + n^2\epsilon_3^2 - (l^2\epsilon_1 + m^2\epsilon_2 + n^2\epsilon_3)^2 \,. \tag{A3.7}$$

Substituting $l = m = n = 1/\sqrt{3}$ into Equations (A3.6) and (A3.7) we obtain

$$\epsilon_{\mathrm{oct}} = \tfrac{1}{3}(\epsilon_1 + \epsilon_2 + \epsilon_3) \tag{A3.8}$$

and

$$\gamma_{\mathrm{oct}}^2 = \tfrac{1}{3}(\epsilon_1^2 + \epsilon_2^2 + \epsilon_3^2) - \tfrac{1}{9}(\epsilon_1 + \epsilon_2 + \epsilon_3)^2$$

or

$$\gamma_{\mathrm{oct}}^2 = \tfrac{1}{9}[(\epsilon_1 - \epsilon_2)^2 + (\epsilon_2 - \epsilon_3)^2 + (\epsilon_3 - \epsilon_1)^2] \tag{A3.9}$$

analogous to Equations (1.9) and (1.11) for the normal and shear stresses on the octahedral plane. In terms of general strain coordinates the octahedral shear strain can be shown to be equal to

$$\gamma_{\mathrm{oct}} = \tfrac{1}{3}[(\epsilon_x - \epsilon_y)^2 + (\epsilon_y - \epsilon_z)^2 + (\epsilon_z - \epsilon_x)^2 + 6(\gamma_{xy}^2 + \gamma_{yz}^2 + \gamma_{zx}^2)]^{1/2} \tag{A3.10}$$

analogous to Equation (A3.5) of this appendix.

3.5 Derivation of expressions for the octahedral shear strain under conditions of uniaxial stress and simple shear

(a) *Uniaxial stress*

For incompressible material, $\epsilon_1 + \epsilon_2 + \epsilon_3 = 0$; in the uniaxial test $\epsilon_2 = \epsilon_3$; combining these two conditions we obtain

$$\epsilon_1 = -2\epsilon_2 = -2\epsilon_3 .$$

Substituting for ϵ_2 and ϵ_3 in Equation (A3.9), we obtain

$$\gamma_{\text{oct}}^2 = \tfrac{1}{9}[(\tfrac{3}{2}\epsilon_1)^2 + (\tfrac{3}{2}\epsilon_1)^2]$$

or

$$\gamma_{\text{oct}} = \frac{1}{3}\frac{\sqrt{18}}{2}\epsilon_1 = \frac{1}{\sqrt{2}}\epsilon_1 . \qquad \text{[Equation (3.15)]}$$

(b) *Simple shear*

Under these conditions the only non-zero strain in Equation (A3.10) is γ_{zx}, and we obtain

$$\gamma_{\text{oct}} = \tfrac{1}{3}\sqrt{6}\gamma_{zx} = \sqrt{\tfrac{2}{3}}\gamma_{zx} . \qquad \text{[Equation (3.17)]}$$

Mass movements in rock and soil masses

Glossary of symbols

a angle between failure plane and horizontal [100]
a_p angle between a possible failure plane and horizontal [116]
c cohesion (as a stress; general symbol) [100]
c' cohesion (as a stress; defined in terms of effective stresses) [121]
c_d developed cohesion (as a stress) [117]
h, h_e, h_p total head [123]; elevation head [123]; pressure head [123]
i slope angle (degrees or radians) [100]
k distance [112]; coefficient [117]
l length of slope or slope section [109]
q_u unconfined compressive strength [103]
p osmotic potential [124]
r radius [121]
s shear strength (as a stress) [116]
s_d shear resistance (as a stress) [116]
u pore pressure [109]; velocity [124]
x horizontal distance [102]
z vertical distance [107]

E_n, E_{n+1} interslice normal stresses [121]
F_s factor of safety [116]
H slope height [101]
H_c critical slope height (without tension cracks) [100]
H'_c critical slope height (with tension cracks) [102]
N total normal force [106]
N_s stability number [110]
P true capillary potential [124]
S shear strength (as a force) [107]
T shear force [107]
W weight of wedge of earth [102]
X_n, X_{n+1} interslice shear stresses [121]
Z depth from ground surface to point of intersection between tension crack and failure plane [102]
Z_0 thickness of tension zone [102]

α angle between failure surface and major principal plane [120]
δ angle of top slope [101]
γ bulk unit weight of earth [100]
γ_w bulk unit weight of water [112]
θ angle between failure surface and horizontal [121]
ϕ angle of internal friction (general symbol) [100]; total potential [124]
ϕ' angle of internal friction (defined in terms of effective stresses [109

ϕ'_p, ϕ'_r peak and residual angles of internal friction [113]
ψ 'capillary' potential [124]
π adhesion potential [124]

4.1 Introduction

The flow of water in stream channels and down hillside slopes is attributable entirely to the pull of the gravity force on the fluid mass. This downslope pull disappears only when the fluid attains a level surface, and it is for this reason that surface water flows towards the sea and lakes. Although this same pull acts also on the solid material of the Earth's crust, there is no *continuous* flow of rock and earth towards areas of a level surface. The reason for this is that earth materials possess internal resistance (shear strength) which acts against the downslope pull of the gravity force. As a result, sloping surfaces can occur in solid masses without any mass movement of the material.

There is, of course, a limit to the steepness and length of these sloping surfaces, depending on the magnitude of the shear strength of the material. Some clays which contain little more strength than a mass of water will only be stable at very low angles; if, for some reason, slopes are produced which are steeper than this critical angle (also termed threshold or limiting angle), a mass movement of the clay will occur comparable to the mass flow of water on sloping surfaces. Some solid rocks, in contrast, possess so much shear strength that high vertical slopes may be stable. On a geologic timescale, however, the strength of a rock mass is reduced by various processes; eventually insufficient strength would exist to support a vertical cliff and mass movements will take place. It is therefore important to appreciate that the stability of an earth slope is, on a geologic timescale, only a temporary phenomenon, and as a result, although there is no continuous flow of earth material downslope as in the case of surface water, there is an intermittent succession of mass movements of earth material occurring in the landscape.

The flow of channel and non-channel water over the ground surface is interesting to earth scientists for two reasons. In the first place, it is responsible for the transport of debris down slopes and is, thus, a major agent of erosion. Secondly, during this process of erosion, it imposes a distinctive form on the landscape; erosion produces new sloping surfaces on which, at any point, the surface slope (together with the characteristic depth of flowing water) is just sufficient to transport water and sediment from upslope, but is insufficiently steep to produce further debris transport from, and lowering of, that surface. In brief, a mass of flowing fluid will tend to attain a definite threshold slope surface.

The intermittent mass movements of earth debris on hillsides are of interest for the same reasons. A landslide often produces erosion at the interface between the sliding mass and the underlying earth in a similar way to the erosive action of flowing water. Note, however, that the material moved by the tractive force of a sliding earth-mass is usually negligible in comparison with the amount of material actually involved in the mass movement itself. In contrast to fluid erosion, therefore, it is the mass movement itself, rather than the secondary erosive action of the movement, which is important. On a long timescale, the effect of these intermittent mass movements must be to produce slopes of lower angles. This may take place through direct flattening (decline hinged about the slope base) or retreat of hillslopes and concomitant emergence of gentler basal slopes. If the shear strength of the slope material and the shear stresses due to gravity acting upon it are constant over a sufficiently long period of time, slopes will become completely stable at the threshold slope angle determined by the stress and strength values. This threshold slope is therefore similar to W.M.Davis' concept of the graded stream channel slope. Similarly, just as Davis believed that the graded stream profile could attain lower angles of slope over geologic time because of *in situ* reduction of particle size, we can appreciate that reduction in the shear strength of hillside material through weathering will result in lower threshold angles during the course of geologic time. Implicitly this, of course, assumes that the base level of erosion is stable.

This treatment of mass movements is restricted to rapid processes. Space precludes discussion of slow mass movements, such as seasonal soil creep, and the reader is referred to the account provided by Carson and Kirkby (1971). We shall firstly examine rapid mass movements on hillsides which are being actively undercut by streams or other mechanisms at the base. A distinction is made between strong intact rock masses (4.2), cohesionless material (4.3), and clay masses (4.4). Finally, mass movements on hillsides protected from basal undercutting are examined (4.5).

4.2 Undercutting in a strong rock mass

As a stream cuts down into a strong, intact rock mass, vertical canyons are produced. The reason that the canyon walls can stay vertical and do not become unstable is that they possess cohesion (see Chapter 3). There is, however, a limit to how deep vertical canyon walls can be cut before the walls will collapse. Various formulae can be worked out for this critical height H_c depending on the assumptions made about the failure surface.

A common assumption (the Culmann method) is that the failure surface is a plane passing through the toe of the slope (Figure 4.1); field observations indicate that this is often a satisfactory assumption for vertical slopes. With the Culmann method it can be shown (Appendix 4.1) that

$$H_c = \frac{2c}{\gamma} \frac{\sin i}{\sin(i-a)(\sin a - \cos a \tan\phi)} \tag{4.1}$$

which, on the assumption that $a = \frac{1}{2}(i+\phi)$, can be reduced (Appendix 4.2) to

$$H_c = \frac{4c}{\gamma} \frac{\sin i \cos\phi}{[1-\cos(i-\phi)]} \tag{4.2}$$

where H_c is the critical height; c the cohesion of the rock; ϕ the angle of internal friction of the rock; γ the total bulk unit weight of the rock; and i the angle of the surface slope.

The equations can be viewed in two ways. For a steeper slope, the critical height will be smaller. Alternatively, as a valley gets deeper, the critical angle of stability becomes smaller. As an illustration, for $i = 70°$ and $\phi = 0$,

$$H_c = \frac{4c}{\gamma} \times 1{\cdot}42\ ;$$

for $i = 20°$ and $\phi = 0$,

$$H_c = \frac{4c}{\gamma} \times 5{\cdot}65\ .$$

These changes are precisely those which would be expected; at higher angles of surface slope i, the potential failure plane is inclined at a higher angle $a\ [= \frac{1}{2}(i+\phi)]$ and therefore a smaller depth of slope is necessary to produce a mass movement.

If the stream is cutting down vertically, initially $i = \pi/2$, and we have

$$H_c = \frac{4c}{\gamma} \frac{\cos\phi}{1-\sin\phi}\ , \tag{4.3}$$

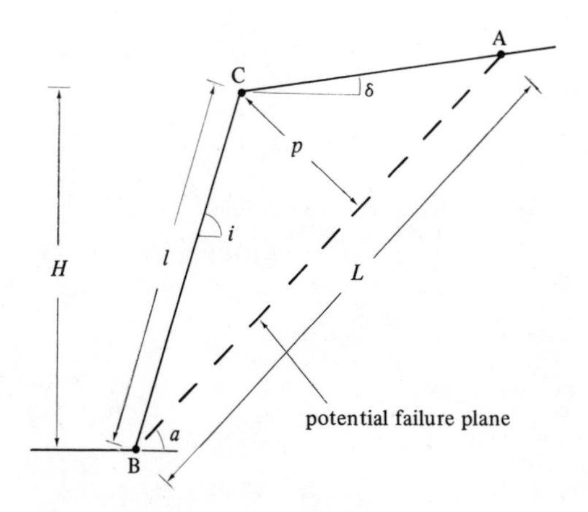

Figure 4.1. The Culmann approach to slope stability.

which is commonly written (Appendix 4.3) in the form

$$H_c = \frac{4c}{\gamma} \tan\left(\frac{\pi}{4} + \frac{\phi}{2}\right) \qquad (4.4)$$

and which for a frictionless rock ($\phi = 0$) reduces to

$$H_c = \frac{4c}{\gamma} . \qquad (4.5)$$

Actually, Equations (4.1)–(4.5) are only valid in special cases. Sometimes before H_c is reached *tension cracks* develop (Figure 4.2) and, if these extend down to the potential failure plane, the critical height is reduced. If the critical height in the case of slopes with tension cracks is denoted by H_c', and the depth of the tension crack above the failure plane by Z, it can be shown (Appendix 4.4) that

$$H_c' = H_c - Z , \qquad (4.6)$$

where H_c is the critical height which could have been attained if tension cracks had not developed prior to failure. Unfortunately, it is very difficult to predict Z because it depends not only on the depth of the crack but also on its position. There is, however, a limit to the depth of the crack; this is the thickness of the surface zone of tension. If the thickness of this zone is denoted by Z_0, it can be shown (Appendix 4.5) that

$$Z_0 = \frac{2c}{\gamma} \tan\left(\frac{\pi}{4} + \frac{\phi}{2}\right) \qquad (4.7)$$

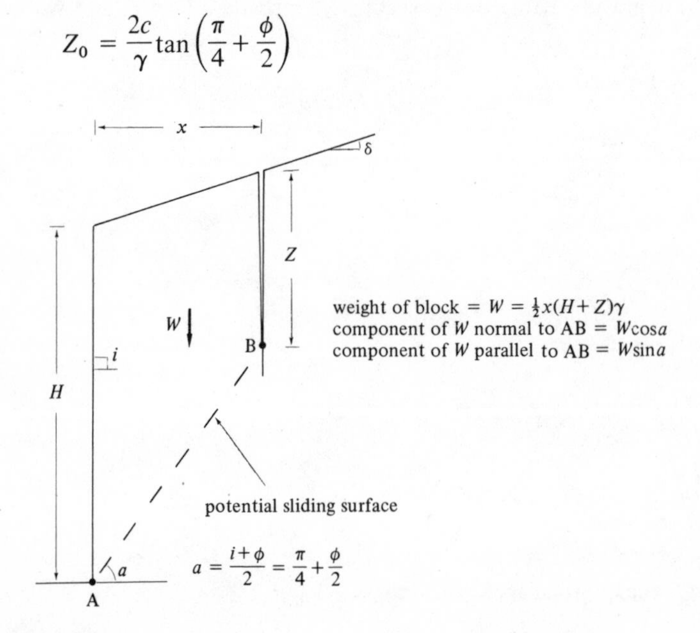

Figure 4.2. The influence of tension cracks on slope stability.

so that, *in the worst possible case* ($Z = Z_0$), the critical height of a vertical cliff is given [combining Equations (4.4), (4.6), and (4.7)] by

$$H_c' = \frac{2c}{\gamma}\tan\left(\frac{\pi}{4}+\frac{\phi}{2}\right) = Z_0 . \tag{4.8}$$

It may be noted that Terzaghi (1943) cites empirical evidence that Z is often approximated by

$$Z = \tfrac{1}{2}H_c' ;$$

and in this case Equation (4.6) yields

$$H_c' = 2\cdot67\frac{c}{\gamma}\tan\left(\frac{\pi}{4}+\frac{\phi}{2}\right) . \tag{4.9}$$

Sometimes these formulae appear to be quite accurate predictors of maximum cliff height. Lohnes and Handy (1968), for instance, working in the löss area of western Iowa, found close agreement between actual and theoretical [based on Equation (4.4)] values for critical height. In this area, the löss had the following properties:

$c = 0\cdot091$ kg cm^{-2} ($1\cdot3$ psi)
$\phi = 25°$
$\gamma = 1208$ kg m^{-3} (76 lb ft^{-3}) .

Substituting these values into Equation (4.4) yields

$$H_c = \frac{4(0\cdot091)\times\tan 57\tfrac{1}{2}°\times 10000}{1208}\ \mathrm{m} \approx 4\cdot7\ \mathrm{m} .$$

Field survey showed the actual heights of vertical cliffs to be, at a maximum, $4\cdot5$ m. Although it was observed that tension cracks do develop in the löss area, because they do not develop until cliffs are already on the verge of failure, there is no need to accommodate them into the slope stability formula.

At other times, as emphasized by Terzaghi (1962), these formulae appear to greatly overestimate the critical height of cliff walls. Consider the case of a medium strength rock with an unconfined compressive strength q_u of about 350 kg cm^{-2} (5000 psi), and a bulk unit weight γ of about 2700 kg m^{-3} (170 lb ft^{-3}). If we use the unconfined compressive strength (Appendix 4.6) rather than the separate strength parameters c and ϕ, Equation (4.8) (the worst case) takes the form:

$$H_c' = \frac{q_u}{\gamma} \tag{4.10}$$

and, with the typical values quoted above, we would obtain

$$H_c' = 1300\ \mathrm{m} .$$

We should emphasize again that this theoretical value is for a medium-strength rock under the worst conditions (zero tensile strength); stronger rocks such as granites or gritstones with q_u values much greater than 350 kg cm^{-2} should be able to stand much higher. Very few vertical canyon walls, however, approach 1000 m in height. We must conclude that laboratory values of unconfined compressive strength are unrealistic. Terzaghi points out that the unconfined strength of a small rock specimen conveys little relevant information about the strength of the rock mass as a whole, and that, in the case of highly fractured rock masses, these rocks may in fact be completely cohesionless. The crux of the problem is this. A highly fractured rock mass contains numerous flaws, cracks, and joints. As a valley is cut into it, the shear stresses that develop are concentrated on these points of weakness and the cracks open up. The stresses must now be taken up somewhere else and a chain reaction develops. Eventually the mass breaks down into a densely-packed aggregate of coarse rock fragments; because it has lost its cohesion it cannot stand vertically [$c = 0$ in Equation (4.3)]. Such rock masses are considered in Section 4.3.

To continue with the development of intact rock slopes let us assume that streams cut down sufficiently deep that a mass movement occurs; if this does not happen, then we can bypass the next part of the sequence and move directly to Section 4.5. A new slope is created by the mass movements and, according to the Culmann approach, it is in fact the failure plane of the mass movement. It was previously noted (Appendix 4.2) that the angle of this plane with the horizontal is given by $a = (i+\phi)/2$; for $i = \pi/2$, $a = \pi/4+\phi/2$. The question that now arises is: What happens if the stream continues cutting down? Two possible models are shown in Figure 4.3. In both schemes, continued downcutting results in repeated intermittent failure with successively lower-angled slopes emerging; but the relationship between the angle of the old failure plane a and the angle of the new surface slope i is different. In model 1 it is assumed that on exposure no erosion takes place on the former failure plane; as the stream cuts down, adding a vertical component to the slope, the effective slope angle i increases. In model 2 it is assumed that on exposure processes such as soil wash lower the old failure plane at the same angle and at the same rate as the stream cuts down. So far there is little evidence to indicate if either of these models actually occurs in nature. In the survey by Lohnes and Handy in the Iowa löss, the authors describe a sequence which combines both models. As we have seen, the first stage of this sequence is the production of a new slope after failure once the cliff reaches about 4·5 m in height. Continued undercutting results in a second failure when the average slope angle i is steepened to about 77°. This is simply predicted (Figure 4.4) from Equation (4.3) after noting that i (average slope angle) increases as the depth of undercutting increases, in accordance with model 1 of Figure 4.3.

According to the Culmann theory, the second failure plane makes an angle $a = \frac{1}{2}(77+25)° = 51°$. This figure agrees with field evidence. Figure 4.5 shows the frequency distribution of slope angles in the löss as indicated by field survey and a distinct peak occurs at 51°. Another peak occurs at 38° and this could be explained in terms of model 2 of Figure 4.3.

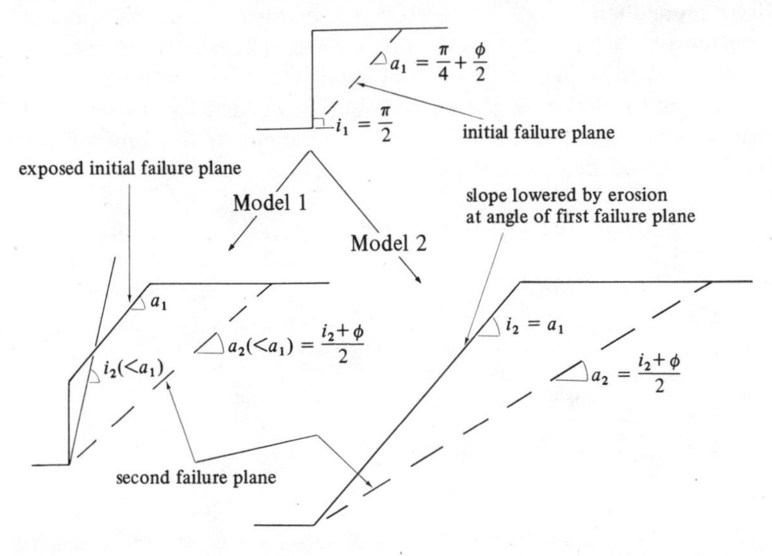

Figure 4.3. Alternative models of slope development during downcutting in a strong rock.

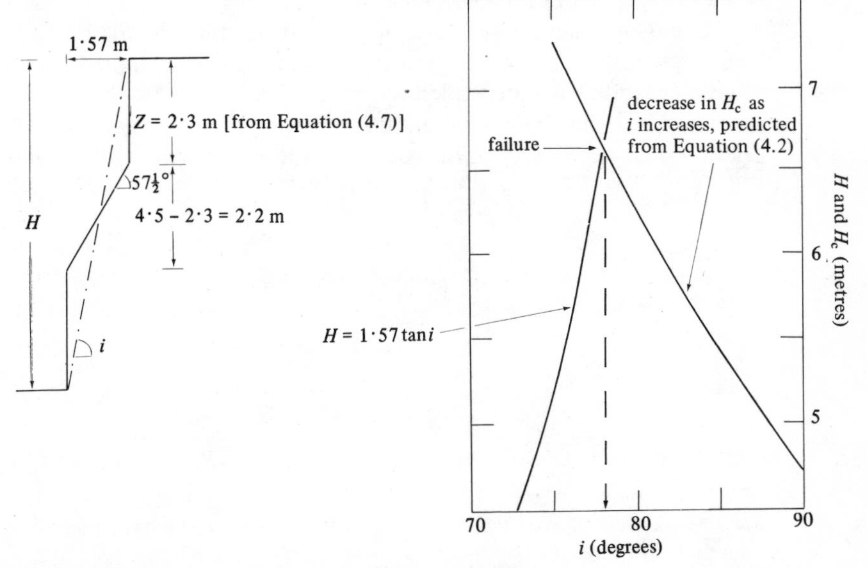

Figure 4.4. Prediction of failure conditions in model 1.

Undercutting of löss slopes at an angle of 51° would result in a third failure along a plane at an angle $a = \frac{1}{2}(51+25)° = 38°$ to the horizontal. This, however, may be purely fortuitous, because the angle of repose of löss in a loose dry state was also found to be 38°.

In most rock masses the models depicted in Figure 4.3 are unlikely to have more than theoretical interest. The reason is the point made previously; by the time a reasonably deep valley has been cut into a rock mass, particularly in fractured rock-masses, the rock is likely to have lost most of its cohesion in the near-surface layers and to behave as a densely-packed mass of rock rubble. The development of this type of slope is more properly dealt with in the next section.

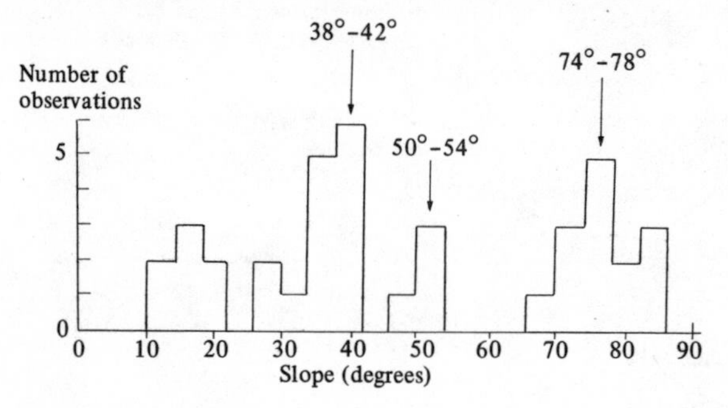

Figure 4.5. Frequency distribution of slope angles in löss, western Iowa (after Lohnes and Handy, 1968).

4.3 Undercutting in a cohesionless rock mass

Slopes cut in cohesionless material differ from those in cohesive rock in two important respects: they cannot stand vertically, and the limiting angle of stability is independent of slope height. In the case of coarse rock material the angle of limiting stability is equal to the angle of internal friction ϕ of the debris. This is easily shown by reference to Figure 4.6.

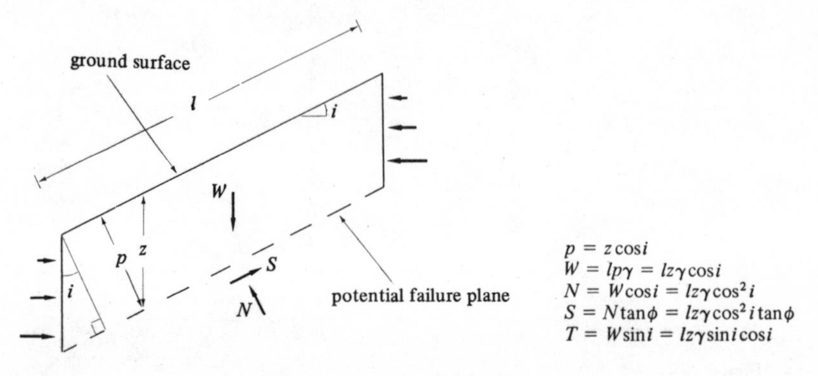

Figure 4.6. Forces in an infinite slope of cohesionless material.

At limiting equilibrium the shear force and shear strength along the potential failure surface are just balanced and are, for any plane parallel to the surface, given by

$$T = Lz\gamma \cos i \sin i \,, \tag{4.11}$$

$$S = Lz\gamma \cos^2 i \tan\phi \,, \tag{4.12}$$

where T is the shear force and S is the shear strength. If we equate these two expressions (for the case of limiting stability) we obtain

$$\tan i = \tan\phi \,. \tag{4.13}$$

The actual stability analysis of slopes formed by undercutting in cohesionless rock material is thus quite simple. Valleys are V-shaped, and side slopes are maintained at an angle $i = \phi$ by shallow mass movements as undercutting progresses. The mechanism involved in this process is shown in Figure 4.7. Assume initially that the slope is in limiting equilibrium, with $i = \phi$. Allow the stream to cut down a very small distance z in the period of time between t_1 and t_2. The steepest potential failure plane that can pass through the toe of the slope is now greater than ϕ. Slides will therefore occur until the slope is reduced to $i = \phi$ again, at which time the steepest potential failure plane passing through the toe is just less than $i = \phi$. The difficult part of applying this model to actual slopes in nature is the determination of ϕ for the material that constitutes the rock mass. In the case of fine material, such as sand-size debris, this is relatively easy. A sample of material is placed in a small direct-shear apparatus or triaxial cell (see Chapter 3), and its strength is determined at different normal pressures. [We have previously emphasized (Chapter 3) that ϕ depends on the density of packing of the material; for sands ϕ may vary between 45°–50° (dense) and 30°–35° (loose) depending

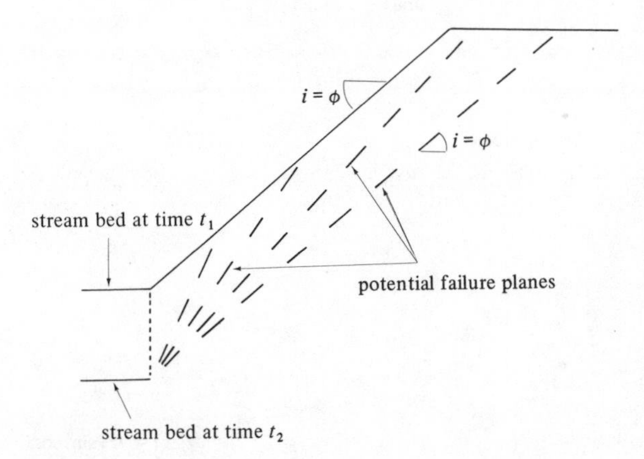

Figure 4.7. Mechanism of maintenance of slope at $i = \phi$ during undercutting of slope base.

on the state of packing.] The task of estimating ϕ is much more difficult for large rock fragments such as those that occur in highly fractured rock masses; these fragments are simply too large to be accommodated in conventional shear apparatus. Tests on material up to 10 cm in size (Silvestri, 1961) have shown, however, that ϕ may be as high as 65° for densely-packed material, whereas in a loose state the same material has a ϕ value close to 35°.

This difference between ϕ for cohesionless material in dense and loose states of packing has important implications for slope development. One of the effects of weathering is to loosen the fragments in a rock mass, particularly near the surface. Because weathering and undercutting occur simultaneously in slopes formed by stream incision into a rock mass, the balance between the two is very important. Under conditions of rapid undercutting, weathering and loosening of the rock material is relatively slow and slopes will stand at an angle equal to ϕ_{dense}; under conditions of very slow downcutting, weathering may become dominant and slopes will stand at an angle equal to ϕ_{loose}. Support for this idea is provided by examining some of the data collected by Strahler (1950) for slopes in the southwestern USA. Figure 4.8 shows the frequency distribution of slope angles (maximum hillside slope angle) in the Verdugo Hills area. Slopes at 43°–45° are essentially cohesionless rock slopes (high density of packing)

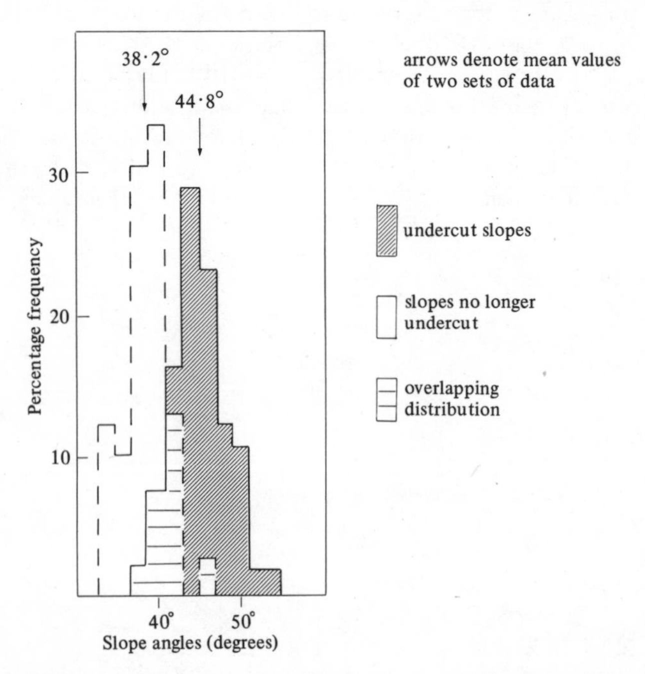

Figure 4.8. Frequency distribution of slope angles in Verdugo Hills, California (after Strahler, 1950).

and occur where stream undercutting is rapid; slopes at 38° are essentially talus slopes, found where slopes are protected from basal undercutting, and standing at the angle of repose (ϕ for loose state of packing) of the rock debris.

Equation (4.13) is, however, only valid under certain conditions; it assumes that in the state of limiting stability the voids, or pore spaces, between rock fragments are dry. This may be true for coarse material which contains pores that are so large that they will not become saturated and after a rain storm will drain completely. As noted in Chapter 3, partially-saturated or fully-saturated material is subject to non-atmospheric pore pressures which exert an appreciable effect on the amount of frictional strength which can be developed. Application of Equation (3.5) to the situation illustrated in Figure 4.6, provides a more general expression for shear strength

$$S = l(\gamma z \cos^2 i - u) \tan \phi' \tag{4.14}$$

which simplifies to Equation (4.12) in the special case when u (the pore pressure) is zero relative to atmospheric pressure. The effect of positive pore pressures is thus to reduce the shear strength and reduce also the angle of limiting stability for cohesionless material to a value given by

$$\tan i = \frac{(\gamma z \cos^2 i - u)}{\gamma z \cos^2 i} \tan \phi' . \tag{4.15}$$

Saturation of the surface weathered mantle is, however, a relatively infrequent occurrence. If streams are cutting down rapidly into a rock mass, most mass movements will be unaffected by the occasional development of excess pore pressures, and as a result the slope angle at limiting stability will be controlled basically by the ϕ value. Under conditions of slow or no downcutting these intermittent periods of saturation become relatively more important. Discussion of their effect on slope instability is deferred until Section 4.5.

4.4 Undercutting in a clay mass

Clays possess cohesion, but only in very small amounts when compared with rocks that have been subjected to high pressures during their formation. Certain aspects of instability in clays therefore resemble instability in strong rocks, whereas other aspects resemble mass movements in cohesionless material.

In the first place, *deep* vertical slopes cannot occur in clay masses. Typical values for the basic engineering parameters of a clay mass, such as the London Clay, are:

$$c = 750\text{–}1250 \text{ kg m}^{-2} ,$$
$$\phi = 10°\text{–}30° ,$$
$$\gamma = 1900\text{–}2250 \text{ kg m}^{-3} .$$

Substitution of average values into Equation (4.4) yields the following value for the critical height of a vertical slope:

$$H_c = \frac{4 \times 1000 \times 1 \cdot 7}{2000} = 3 \cdot 5 \text{ m} ,$$

As streams cut down to depths greater than this critical value, application of Equation (4.2) indicates that we should expect a continual, though intermittent, decrease in slope angle as slope height (valley depth) increases. Actually, it is misleading to invoke the Culmann method in such a sequence of downcutting because deep slips will, at least after the first one, tend to be rotational rather than planar. This curvature of the failure plane follows directly from the Mohr circle diagram (Figure 3.17) for a sloping mass in a cohesive material. The effect of this is to *slightly* reduce the value of H_c given by Equation (4.2).

Some indication of the relationship between slope height and slope angle in the limiting state is provided by the stability charts (Figure 4.9) constructed by Taylor (1948). These may be used in one of two ways. For a given slope angle the stability number ($N_s = \gamma H/c$) can be read off the chart for a particular ϕ value. From a knowledge of the bulk unit weight and the cohesion of the clay, the critical height H_c is easily determined. Alternatively, for any given set of c, ϕ, and γ values, we can compute the angle of limiting slope for any specified valley depth H. As an example, consider a clay with properties $c = 1000$ kg m^{-2}, $\gamma = 2000$ kg m^{-3}, and $\phi = 20°$. For a valley 50 m deep, $N_s = 100$. A horizontal line from $N_s = 100$ intersects the $\phi = 20°$ curve at $22\frac{1}{2}°$,

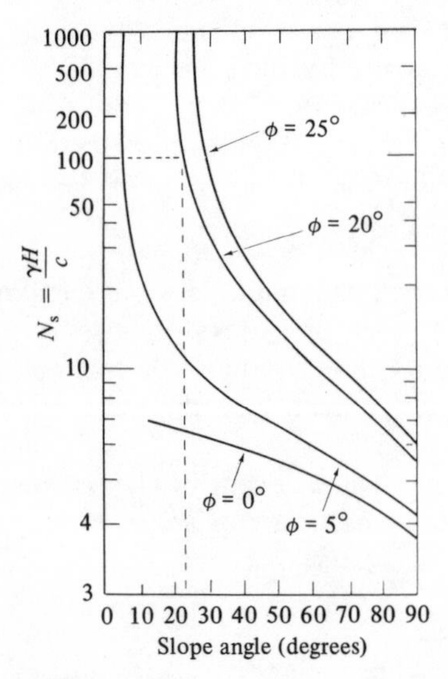

Figure 4.9. Stability chart for critical toe circles (after Scott, 1963).

which is the limiting angle of stability. Note that, irrespective of the valley depth H (and therefore the N_s value), the angle of limiting stability will not be less than ϕ.

Figure 4.9 indicates, in the same way as Equation (4.2), that the angle of limiting stability decreases as slope height (expressed in N_s) increases. Note, however, that because the γ/c ratio for clays is so large (typically about 2, as in the example) relatively small values of H are needed to produce high N_s values. Once slope heights of about 50 m are attained (N_s = 100), the rate of decrease of limiting slope angle with increasing H is very small. Valley-side slopes in clay masses will thus be very close to the ϕ value once depths of about 50 m have been attained and will change very little as the valley deepens. In this respect, slopes in clay masses may be treated as slopes formed in cohesionless material.

So far we have not given the details of a stability analysis for a rotational slip; we have merely cited the analogy with the formula for the Culmann method. There are in fact a number of different methods available, although the most commonly used approach is that of Bishop (1955). The details of this method are presented in Appendix 4.7. The relationship of this method to the stability analysis of an infinite slope subjected to planar slides [Equation (4.13)] is discussed in Appendix 4.8. Before leaving the topic of undercutting in clay slopes, we should emphasize that, as in the previous two sections, it has been assumed that excess pore pressures do not occur in these instability problems. If downcutting is slow this may be invalid, and for this reason, Bishop and Morgenstern (1960) have expanded the stability charts of Taylor to accommodate various pore pressure conditions. This problem of pore pressures is dealt with below.

4.5 Instability without undercutting

Throughout the previous discussion we have dealt with instability produced by undercutting of slopes. In the case of cohesionless material this results (Figure 4.7) from fractionally steepening the average slope above ϕ. In the case of cohesive rocks (Figure 4.1) this stems from increasing the height of slope and thus the weight of material above the potential failure plane; instability occurs because, for cohesive rocks, the shear stress increases more rapidly than the shear strength along the potential failure surface as slope height increases. We shall now consider the case when active undercutting of slopes has stopped and that weathering is the most important process acting on the slopes. As weathering proceeds, the strength of the hillside material decreases over time, and we may expect further mass movements, even though the shear stresses inside the hillmass change little. Before dealing with the effect of weathering on shear strength let us first examine the influence of intermittent saturation and the development of positive pore pressures on slope stability.

As a starting point consider a clay slope standing at the angle of internal friction of the clay. Clay minerals are often the last products of weathering and for this reason we may assume that further weathering of an existing clay mass, and thus alteration of c and ϕ, is a relatively slow process. This allows us to concentrate on the role of pore pressures. As noted in Equation (4.15), the development of positive pore pressures in surface soils by a rise in the water table results in a decrease in the angle of limiting stability below ϕ. A simple case is shown in Figure 4.10. The water table has risen to the surface of the hillside and the flow of ground water is assumed to be parallel to the surface slope. It can be shown from elementary mechanics that the flow lines in isotropic material (permeability constant in all directions) are normal to a second set of lines termed *equipotentials*. [An equipotential is a line joining points of equal head or equal potential (Appendix 4.9). The two major components of head are positional head and pressure head (see Figure 4.10). Along any given equipotential the decrease in positional head with depth relative to an arbitrary datum is just balanced by an increase in pressure head.] This has very important implications for slope stability studies. From a knowledge of the groundwater flow pattern and of the degree of anisotropy in the permeability of the hillside material the pattern of equipotentials can be constructed; from this pattern the relationship between pore pressure and depth can be obtained, and by substitution for u in Equation (4.15) the effect on shear strength can be ascertained. Let us return to the particular pattern of groundwater flow in Figure 4.10. The pore pressure at any depth z is given by

$$u = \gamma_w z \cos^2 i \,. \tag{4.16}$$

Substitution for u in Equation (4.15) yields

$$\tan i = \frac{\gamma - \gamma_w}{\gamma} \tan\phi' \,. \tag{4.17}$$

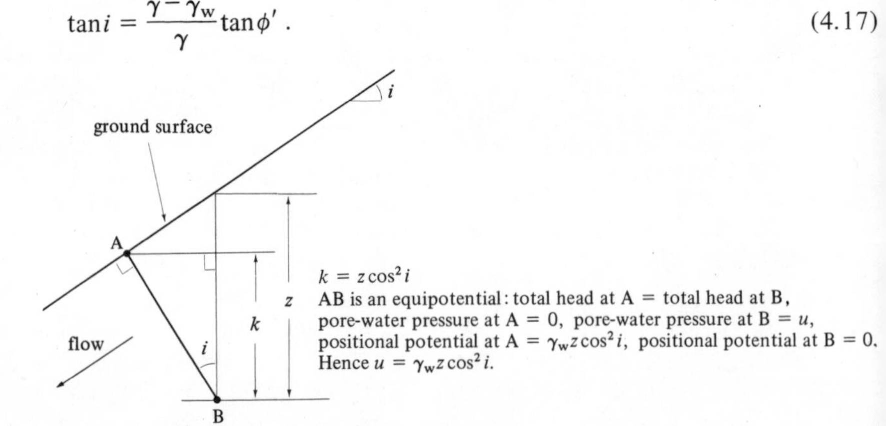

Figure 4.10. The relation between pore-water pressure and depth on a hillside with groundwater flow parallel to the surface.

The terms γ_w and γ are not fixed. Table 2.1 provides values of γ_w ranging from 1000 kg m^{-3} (62·4 lb ft^{-3}) at 0°C (32°F) to 997 kg m^{-3} (62·2 lb ft^{-3}) at 27°C (80°F). The bulk unit weight of saturated soils may vary between 1600 kg m^{-3} (100 lb ft^{-3}) and 2240 kg m^{-3} (140 lb ft^{-3}); it is thus approximately twice the unit weight of water. Equation (4.17) may thus be simplified to

$$\tan i \approx \tfrac{1}{2}\tan\phi' \,. \tag{4.18}$$

This equation (4.18) has been used on several occasions in the long-term stability analysis of natural hillslopes. The reason is that in many cases it may be regarded as the worst groundwater pattern that will affect hillsides. These conditions may occur only infrequently, with return periods of more than a hundred years, but on a geologic timescale they will control the ultimate angle of limiting stability. Skempton and Delory (1957) noted that the ultimate angle of limiting stability in London clay, about 10°, could be explained in this way: measured values of ϕ_r' for the surface layers of London clay are about 16° (Skempton, 1964) and use of Equation (4.18) yields a value of i close to 10°. Although the assumption that the water table can rise to the surface of hillsides may be valid for gentle slopes, it seems unreasonable for steep hillsides. Under certain circumstances Equation (4.16) may, however, be valid on steep slopes. Many hillsides are not homogeneous, but are basically rock masses mantled by more permeable weathered material. Percolation of water into the hillmass from the surface during prolonged rainstorms will be retarded at the contact zone between the more permeable soil and the less permeable rock. Under these conditions the soil mantle may become saturated with freely-draining water and a perched groundwater system, with flow parallel to the surface slope, may occur. Equation (4.16) is valid in this case also. Note that this is only applicable to situations where the soil mantle is more permeable than the underlying bedrock; it would be inappropriate for a clay mantle derived from the weathering of a shale rock mass.

The change from a stable slope at $i = \phi$ to another slope given by $\tan i = \tfrac{1}{2}\tan\phi'$, described for a clay mass above, may be regarded as a one-phase instability sequence. Materials which undergo appreciable weathering may experience more than one phase of instability. Consider a talus slope, standing at $i = \phi = 35°$, derived from the weathering of a highly-fractured rock mass. If undercutting of the slope ceases, the hillside will temporarily remain stable at this angle; voids between rock fragments are sufficiently large that pore pressures can be assumed to be atmospheric. Further weathering will convert some fragments into soilsize debris and produce filling in of the voids. This has two effects. Firstly, interlocking strength is increased; typical ϕ values for *taluvial* material (talus and colluvial material) in a loose state of packing are about 10° higher than ϕ values for talus. Tests by engineers indicate that ϕ values in the range 43°–46° are very common for this type of material irrespective of the actual rock type. Secondly, the reduced permeability associated with the filling in of voids means that the assumption that positive pore pressures do not occur during prolonged rainstorms is now invalid. The net effect of these two changes is to produce further instability; application of Equation (4.18), for instance with $\phi' = 45°$, indicates an angle of limiting stability of about 25°. Many workers have reported slopes, mantled by taluvial material, at angles of close to 25° (Ruxton, 1958; Young, 1961; Melton, 1965; Robinson, 1966) and this type of stability analysis suggests an explanation for the observation. Slopes at the angle of limiting stability for taluvium are, however, only temporary features. Further weathering of the hillside will convert the entire mantle to colluvium. This breakdown of the rock fragments results in a reduction of interlocking strength and lower ϕ' values for colluvial material. Sandy soil mantles in a loose state of packing show ϕ' values of about 35°; for clay soils the friction parameter is smaller. The ultimate angle of limiting stability is therefore closely related to the final weathering product as well as to the hydrologic regime of the hillside.

The number of phases of instability through which a hillside passes, once undercutting at the base comes to a halt, depends on a number of factors. Firstly, it depends on the stage of weathering of the rock mass just prior to the halt in undercutting. If the previous phase of undercutting was rapid, the initial slope might have been a rock slope ($i = \phi_{dense}$) rather than a talus slope ($i = \phi_{loose}$) as pointed out in Section 4.3. In that case, an extra phase of instability would occur. Secondly, it depends on the complexity of the weathering process. Some rocks (e.g. a clay mass) may weather quickly into the final weathering products, whereas others may pass through several stages. Thirdly, it depends on the climatic setting. This may terminate the weathering sequence prematurely. In semi-arid areas, for instance, the fine material derived from the weathering of talus is often washed off the talus slope almost as soon as it is produced. The weathering sequence in these conditions is thus halted at the talus stage.

4.6 End-note

The application of elementary mechanics to mass movements on rock- and soil-mantled hillside slopes should enable us to interpret hillslope development under those processes more successfully than has been the case in the past. There are, however, many problems involved in this work. It is very difficult, for instance, to assess ϕ for rock material that comprises very large fragments. In the previous discussion we have merely indicated that, in a dense state of packing, ϕ may range between 45° and 65°, a rather wide range. A second problem is the choice of the appropriate pore pressure values to use in long-term stability analyses. It is difficult enough to measure the current pore pressure distribution within a hillside; it is impossible to establish with certainty the appropriate pattern on a geologic timescale. And, lastly, as indicated in Chapter 3, there is the problem that clay soils do not have a unique strength value. There is thus no simple distinction between stable and unstable for clay slopes. Much depends on the time scale involved.

Bibliography

Baver, L. D., 1956, *Soil Physics* (John Wiley, New York).
Bishop, A. W., 1955, "The use of the slip circle in the stability analysis of slopes", *Géotechnique*, **5**, 7–17.
Bishop, A. W., Morgenstern, N. R., 1960, "Stability coefficients for earth slopes", *Géotechnique,* **10**, 129–150.
Carson, M. A., 1969, "Models of hillslope development under mass failure", *Geographical Analysis,* **1**, 76–100.
Carson, M. A., Kirkby, M. J., 1971, *Hillslope Form and Process* (Cambridge University Press, Cambridge).
Carson, M. A., Petley, D. J., 1970, "The existence of threshold hillslopes in the denudation of the landscape", *Trans. Inst. Br. Geogr.,* **49**, 71–95.
Hutchinson, J. N., 1969, "A reconsideration of the coastal landslides at Folkestone Warren, Kent", *Géotechnique,* **19**, 6–38.
Lohnes, R. A., Handy, R. L., 1968, "Slope angles in friable loess", *J. Geol.,* **76**, 247–258.
May, D. R., Brahtz, J. H. A., 1936, "Proposed methods of calculating the stability of earth dams", *Transactions of the 2nd Congress on Large Dams,* **4**, 539.
Melton, M. A., 1965, "Debris-covered hillslopes of the Southern Arizona Desert—consideration of their stability and sediment contribution", *J. Geol.,* **73**, 715–729.
Robinson, G., 1966, "Some residual hillslopes in the Great Fish River Basin, S.Africa", *Geogr. J.,* **132**, 386–390.
Ruxton, B. P., 1958, "Weathering and subsurface erosion in granite at the Piedmont angle, Balos, Sudan", *Geol. Mag.,* **95**, 353–377.
Scott, R. F., 1963, *Principles of Soil Mechanics* (Addison-Wesley, Reading, Mass.).
Silvestri, T., 1961, "Determinazione sperimentale de resistenze meccanica del materiale constituente il corpo di una diga del tipo 'rockfill' ", *Geotechnica,* 8, 186–191.
Skempton, A. W., 1964, "Long-term stability of clay slopes", *Géotechnique*, **14**, 75–102.
Skempton, A. W., Delory, F. A., 1957, "Stability of natural slopes in London clay", *Proceedings of 4th International Conference on Soil Mechanics and Foundation Engineering,* Vol.2, pp.378–381.
Skempton, A. W., Hutchinson, J. N., 1969, "Stability of natural slopes and embankment sections", in *Proceedings of 7th International Conference on Soil Mechanics and Foundation Engineering, State of the Art Volume,* pp.291–340.
Strahler, A. N., 1950, "Equilibrium theory of erosional slopes approached by frequency distribution analysis", *Am. J. Sci.,* **248**, 673–696, 800–814.
Taylor, D. W., 1948, *Fundamentals of Soil Mechanics* (John Wiley, New York).
Terzaghi, K., 1943, *Theoretical Soil Mechanics* (John Wiley, New York).
Terzaghi, K., 1962, "Stability of steep slopes on hard unweathered rock", *Géotechnique,* **12**, 251–270.
Vargas, M., Pichler, E., 1957, "Residual soil and rock slides in Brazil", *Proceedings of 4th International Conference on Soil Mechanics and Foundation Engineering,* Vol.2, pp.394–398.
Young, A., 1961, "Characteristic and limiting slope angles", *Z. Geomorphol., Suppbd.,* **5**, 17–27.

Appendix 4

4.1 Derivation of Equation (4.1),

$$H_c = \frac{2c}{\gamma} \frac{\sin i}{\sin(i-a)(\sin a - \cos a \tan\phi)} .$$

The shear force on AB (Figure 4.1) is given by

$$T = W \sin a ,$$

where W is the weight of the wedge ABC; the shear strength along AB is given by

$$S = cL + W \cos a \tan\phi .$$

At limiting equilibrium, the shear force and shear strength are just balanced:

$$W \sin a = cL + W \cos a \tan\phi$$

or

$$W(\sin a - \cos a \tan\phi) = cL . \quad \text{(A4.1)}$$

Now, $W = \frac{1}{2} pL\gamma$, $p = l \sin(i-a)$ and $l = H/\sin i$, so that

$$W = \frac{\frac{1}{2} H \sin(i-a) L\gamma}{\sin i} . \quad \text{(A4.2)}$$

Combining (A4.1) and (A4.2), we obtain

$$\frac{\frac{1}{2} H \sin(i-a) L\gamma(\sin a - \cos a \tan\phi)}{\sin i} = cL$$

which, with $H = H_c$ for limiting stability, reduces to Equation (4.1).

4.2 (a) Derivation of the relationship

$$a = \frac{i+\phi}{2} .$$

In Figure 4.1, the actual shear resistance mobilized along AB, in the *general* condition, is given as a stress by

$$s_d = \frac{c}{F_s} + \frac{\sigma}{F_s} \tan\phi ,$$

where F_s is the factor of safety (at limiting equilibrium F_s equals unity). Consider various possible failure planes passing through B; denote the angle of the *potential* failure planes with the horizontal by a_p. This angle can vary between almost 0° and a maximum angle which coincides with limiting equilibrium. As a_p is increased, F_s decreases, and therefore s_d increases to a maximum value when $F_s = 1$. Thus, the shear resistance along any potential failure plane is up to a maximum a function of the angle a_p. This statement is true for the individual strength parameters c

and ϕ as well as the total resistance s_d. Thus, for cohesion

$$c_d = f(a_p) ,$$

where c_d is the amount of cohesion actually developed on the potential failure plane. If we can determine the function dc_d/da_p and put it equal to zero, we shall from elementary calculus obtain the value of a_p for which c_d is a maximum. This is the angle a.

In Equation (4.1), replace c by c_d, a by a_p, and H_c by H; this yields

$$c_d = k\sin(i-a_p)(\sin a_p - \cos a_p \tan\phi) ,$$

where $k = \frac{1}{2}H\gamma/\sin i$. Differentiating, we have:

$$\frac{dc_d}{da_p} = \frac{d}{da_p}[k\sin(i-a_p)(\sin a_p - \cos a_p \tan\phi)] ,$$

in which the right hand side can be expanded to

$$\frac{d}{da_p}[k(\sin i\cos a_p \sin a_p - \sin i\tan\phi\cos^2 a_p - \cos i\sin^2 a_p + \cos i\tan\phi\sin a_p \cos a_p)]$$

and we obtain

$$\frac{dc_d}{da_p} = k[(\sin i+\cos i\tan\phi)(\cos^2 a_p - \sin^2 a_p) + 2\sin a_p \cos a_p(\sin i\tan\phi - \cos i)] .$$

Putting $dc_d/da_p = 0$ and denoting a_p by a, we have

$$\sin 2a(\cos i - \sin i\tan\phi) = \cos 2a(\sin i + \cos i\tan\phi) ,$$

or

$$\tan 2a = \frac{\sin i + \cos i\tan\phi}{\cos i - \sin i\tan\phi} ,$$

$$= \tan(i+\phi)$$

and

$$a = \frac{i+\phi}{2} . \qquad (A4.3)$$

Note the following point. In the case when $i = \pi/2$, Equation (A4.3) yields

$$a = \pi/4 + \phi/2 ,$$

but from Chapter 3 we know that this is the angle that the failure surface makes with the major principal plane. The implication from this is that in the Culmann analysis of a vertical slope the major principal plane is horizontal. The Culmann method thus implicitly assumes an active

Rankine state of stress. Terzaghi (1943, pp.152–153) argues that this assumption is probably not completely valid; as a corollary, the notion of a wedge failing on a *planar* surface is not completely sound even in the case of a vertical bank, although the amount of error introduced is very small. Once $i < \pi/2$, the concept of a planar slide becomes much less tenable, as pointed out in Section 4.4.

(b) Proof, based on Equation (A4.3), of the relationship

$$H_c = \frac{4c}{\gamma} \frac{\sin i \cos\phi}{[1-\cos(i-\phi)]} . \qquad \text{[Equation (4.2)]} .$$

Substituting for $a\, [= (i+\phi)/2]$ in Equation (4.1), we obtain:

$$H_c = \frac{2c}{\gamma} \frac{\sin i \cos\phi}{\sin[(i-\phi)/2]\{\sin[(i+\phi)/2]\cos\phi - \cos[(i+\phi)/2]\sin\phi\}}$$

whence, using the formula $\sin a \cos b - \cos a \sin b = \sin(a-b)$, we find

$$H_c = \frac{2c}{\gamma} \frac{\sin i \cos\phi}{\sin[(i-\phi)/2]\sin[(i-\phi)/2]}$$

which can be further modified using the formula $2\sin^2\theta = 1-\cos 2\theta$:

$$H_c = \frac{4c}{\gamma} \frac{\sin i \cos\phi}{1-\cos(i-\phi)} .$$

4.3 Proof that the following two formulae are identical

$$H_c = \frac{4c}{\gamma} \frac{\cos\phi}{(1-\sin\phi)} \qquad \text{[Equation (4.3)]} ,$$

$$H_c = \frac{4c}{\gamma} \tan(\pi/4+\phi/2) \qquad \text{[Equation (4.4)]} .$$

$$\tan^2(\pi/4+\phi/2) = \frac{[1+\tan(\phi/2)]^2}{[1-\tan(\phi/2)]^2}$$

$$= \frac{[\cos(\phi/2)+\sin(\phi/2)]^2}{[\cos(\phi/2)-\sin(\phi/2)]^2}$$

$$= \frac{1+\sin\phi}{1-\sin\phi}$$

$$= \frac{1-\sin^2\phi}{(1-\sin\phi)^2} ,$$

or

$$\tan^2(\pi/4+\phi/2) = \frac{\cos^2\phi}{(1-\sin^2\phi)^2} .$$

Equations (4.3) and (4.4) are thus identical.

4.4 Proof that

$$H_c' = H_c - Z \qquad \text{[Equation (4.6)]} .$$

Referring to Figure 4.2, at limiting equilibrium, we see that the shear force along AB is just balanced by the shear strength:

$$W\sin a = c\mathrm{AB} + W\cos a \tan\phi , \tag{A4.4}$$

where W is the weight of earth above AB. From the diagram,

$$W = \tfrac{1}{2}x(H+Z)\gamma . \tag{A4.5}$$

Now, Equations (4.1) and (4.2) indicate that the critical height of the cut is uninfluenced by the angle δ of the top of the mass, so that for convenience we may take $\delta = 0$. In this case we have

$$x = (H-Z)\cot a$$

and

$$\mathrm{AB} = \frac{H-Z}{\sin a} .$$

Substituting these values into (A4.4) and (A4.5) we obtain

$$W\sin a = \frac{c(H-Z)}{\sin a} + W\cos a \tan\phi \tag{A4.6}$$

and

$$W = \tfrac{1}{2}(H-Z)(H+Z)\gamma\cot a ; \tag{A4.7}$$

combining these two equations yields:

$$\tfrac{1}{2}(H-Z)(H+Z)\gamma\cot a(\sin a - \cos a\tan\phi) = \frac{c(H-Z)}{\sin a}$$

or

$$H_c' + Z = \frac{2c}{\gamma}\frac{1}{\cos a(\sin a - \cos a \tan\phi)}$$

where H_c' indicates the critical height after a tension crack has developed prior to failure. Comparison with Equation (4.1), for the case in which $i = 90°$, indicates that

$$H_c' + Z = H_c$$

where the presence or absence of the prime indicates, respectively, the critical height with or without tension cracks.

4.5 Proof that

$$Z_0 = \frac{2c}{\gamma}\tan(\pi/4+\phi/2) \qquad \text{[Equation (4.7)]}.$$

This is easily shown by reference to Mohr's circle of stress for a cohesive material (Figure A4.1) at the depth where the lateral stress σ_3 is zero. At this depth Z_0 the vertical stress is given by

$$\sigma_1 = \gamma Z_0$$

assuming the active state of stress. In addition we have from the diagram

$$\sigma_1 = 2r$$

so that

$$Z_0 = \frac{2r}{\gamma} = \frac{2}{\gamma}c\tan\alpha .$$

The angle α is the angle between the failure surface and the major principal plane; in active failure the major principal plane is horizontal. Thus the angles α and a (Figure 4.2) are, in this case, identical and, with $i = \pi/2$, we obtain $\alpha = \pi/4+\phi/2$. The depth of tension is thus given by

$$Z_0 = \frac{2c}{\gamma}\tan(\pi/4+\phi/2) .$$

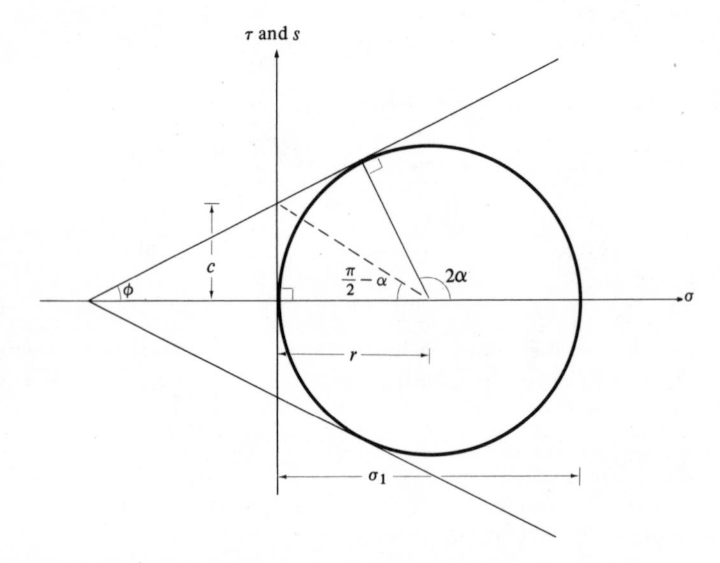

Figure A4.1. The Mohr circle of stress for $\sigma_3 = 0$.

4.6 Proof that in the worst stability case the critical height of a vertical slope is given by

$$H_c' = \frac{q_u}{\gamma},$$

where q_u is the unconfined compressive strength of the slope material.

From Equations (4.4), (4.6), and (4.7), it follows immediately that

$$H_c' = \frac{2c}{\gamma}\tan(\pi/4+\phi/2).$$

Now, refer to the Mohr circle of Figure A4.1. This depicts the state of stress during an unconfined compressive test at failure: $\sigma_3 = 0$ and σ_1 is such that the circle touches the Coulomb strength envelope. From the diagram it is evident that

$$q_u = \sigma_1 = 2r = 2c\tan(\pi/4+\phi/2).$$

4.7 Equations for circular arc stability analyses of clay slopes: the method of slices. There are two main variants of this approach. The conventional method of slices, first proposed by May and Brahtz (1936) is represented by

$$F_s = \sum_A^B \frac{c'l+(W\cos\theta-ul)\tan\phi'}{W\sin\theta}, \tag{A4.8}$$

where F_s is the factor of safety of the slope in Figure A4.2. A modified version of the method of slices was presented by Bishop (1955) and in a simplified form is given by

$$F_s = \sum_A^B \frac{c'l+[(W/\cos\theta)-ul]\tan\phi'}{1+(\tan\theta\tan\phi'/F_s)}\frac{1}{W\sin\theta}; \tag{A4.9}$$

this is derived below.

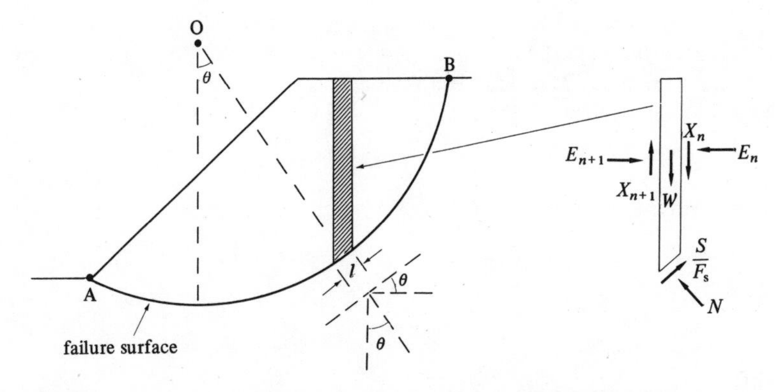

Figure A4.2. Stability analysis of circular slips: method of slices.

The factor of safety of the slope in Figure A4.2 is defined as

$$F_s = \sum_{A}^{B} \frac{S}{T} ,$$

where S and T are, respectively, the maximum shear resistance and the actual shear force along the base of a slice, and are given by

$$S = c'l + (N - ul)\tan\phi' , \qquad \text{(A4.10)}$$

$$T = W\sin\theta .$$

N is the total normal force acting on the base of the slice; it may be determined by resolving forces vertically. At equilibrium

$$N\cos\theta = W + X_n - X_{n+1} - \frac{S}{F_s}\sin\theta \qquad \text{(A4.11)}$$

in which $F_s = 1$ for the case when the slope is on the verge of instability. Substituting for N in the expression for S above, we obtain

$$S = c'l + \left(\frac{W^*}{\cos\theta} - \frac{S\tan\theta}{F_s} - ul\right)\tan\phi'$$

where $W^* = W + X_n - X_{n+1}$. After rearranging the expression simplifies to

$$S = \frac{c'l + [(W^*/\cos\theta) - ul]\tan\phi'}{1 + (\tan\theta\tan\phi'/F_s)} .$$

The factor of safety is thus given by

$$F_s = \sum_{A}^{B} \frac{c'l + [(W^*/\cos\theta) - ul]\tan\phi'}{1 + (\tan\theta\tan\phi'/F_s)} \frac{1}{W\sin\theta} .$$

Bishop (1955) has shown that, with an estimated loss of accuracy of less than 1%, it may be assumed that $X_n = X_{n+1}$, or $W^* = W$, and the expression is then identical to Equation (A4.9). Note, finally, that the difference between (A4.8) and (A4.9) arises in the derivation of the normal force N. May and Brahtz neglected the term F_s in Equation (A4.11), implicitly assuming $F_s = 1$. Equation (A4.11) then becomes

$$N\cos\theta = W^* - S\sin\theta ;$$

moreover, if $F_s = 1$, $S = T = W\sin\theta$, and we obtain

$$N\cos\theta = W(1 - \sin^2\theta) ,$$

or

$$N = W\cos\theta$$

if inter-slice pressures are neglected. If this value of N is inserted into Equation (A4.10) this yields Equation (A4.8) rather than Equation (A4.9).

4.8 The relationship between the stability analyses for rotational slips and for planar slides on infinite slopes

In the stability analysis for an infinite slope subject to planar slides (Figure 4.6) the angle θ (= i) is constant for all slices. We may therefore consider just one slice and ignore the summation sign in Equation (A4.9), which then becomes

$$1 = \frac{c'l + [(W/\cos\theta) - ul]\tan\phi'}{F_s + \tan\theta\tan\phi'}\,\frac{1}{W\sin\theta}$$

or

$$F_s = \frac{c'l + [(W/\cos\theta) - ul]\tan\phi'}{W\sin\theta} - \tan\theta\tan\phi' = \frac{c'l + (W\cos\theta - ul)\tan\phi'}{W\sin\theta},$$

which is identical to the solution (A4.8) proposed by May and Brahtz.

Now, in the situation given by Figure 4.6,

$$W = \gamma z l \cos\theta ,$$

so that for the planar case

$$F_s = \frac{c' + (\gamma z\cos^2\theta - u)\tan\phi'}{\gamma z\sin\theta\cos\theta} . \qquad \text{(A4.12)}$$

For cohesionless soils at limiting equilibrium ($F_s = 1$) Equation (A4.12) reduces to

$$\tan\theta = \frac{\gamma z\cos^2\theta - u}{\gamma z\cos^2\theta}\tan\phi'$$

which, with $i = \theta$, is identical to Equation (4.15).

4.9 Head and potential

A corollary of Darcy's well-known law for the flow of water through a saturated soil is that, at any point, the direction of flow follows the direction of the maximum *hydraulic gradient* at that point, assuming that permeability is constant in all directions. The meaning of hydraulic gradient in saturated soils is quite simple and may be understood by reference to Figure A4.3. If pipes are inserted at two points separated by a distance l, and the difference in elevation between the top of the water in the two pipes is Δh, the hydraulic gradient between the two points is equal to $\Delta h/l$.

The term Δh is the difference in *total head* between the two points. Total head is shown to be the sum of elevation head and pressure head:

$$h = h_e + h_p . \qquad \text{(A4.13)}$$

Head is in fact essentially a measure of energy; by reference to Appendix 2.7 it is seen that the components of head are also found in the Bernoulli equation. Strictly speaking, the total head should also include the velocity head $u^2/2g$, but in seepage problems this is so small that it is usually assumed to be zero.

In *un*saturated soils pressures are negative relative to atmospheric pressure and the term pressure head is usually replaced by capillary potential. Indeed, with reference to unsaturated soils the term potential is generally preferred to the term head; Equation (A4.13) is usually written in the form:

$$\phi = \rho g h + \psi \,, \tag{A4.14}$$

where ϕ denotes total potential, $\rho g h$ is the elevation potential (or positional potential), and ψ is the capillary potential. Note that the dimensions of the components of Equation (A4.14) are those of pressure (weight/area), rather than head (length), as previously. Actually, in the case of unsaturated soils the parameter ψ is really a mask for many effects, not merely capillary potential. It includes at least three components:

$$\psi = P - p - \pi$$

(using the notation of Baver, 1956, p.239), in which P is the true capillary potential, p is the osmotic potential (due to the presence of soluble salts or adsorbed ions), and π is the adhesion potential associated with the attraction of dipolar water molecules with surfaces of particles. If osmotic and adhesion forces are small, the total potential is essentially a function of elevation and pressure.

Lines joining points with equal potential (or head) are called equipotentials. The maximum hydraulic gradient at any point is clearly normal to the equipotentials; thus in isotropic material flow lines are normal to the equipotentials. If the flow lines pass through material of varying permeability, the situation is, however, rather more complicated.

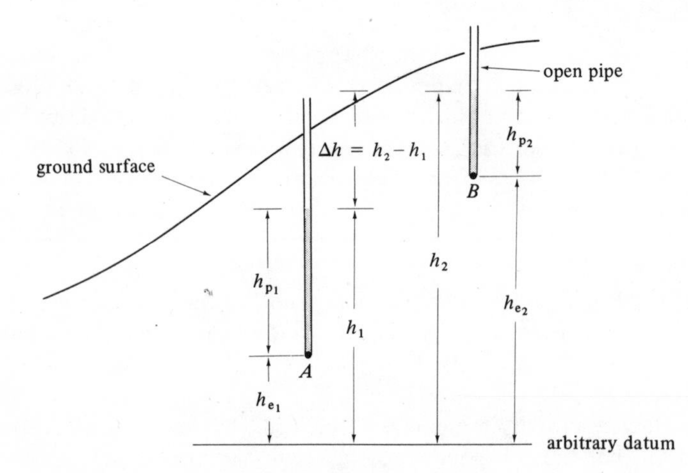

Figure A4.3. Components of hydraulic head.

5

Mechanics of glacial erosion

Glossary of symbols

a height of obstacles on glacier bed [135]; fraction of basal shear surface occupied by sole–rock contact [150]
f basal shear stress [134]
g acceleration due to gravity [128]
h depth of glacier [162]
i slope angle [128]
k, k' coefficients [133]
l length [154]
l_0 initial length [154]
n exponent in flow law [128]
p water pressure [135]
δp pressure difference [131]
r bed roughness ($= a/\lambda$) [135]
s measure of cavitation [162]
t time (duration) [156]
δt incremental time [133]
u velocity [128]
u_s surface velocity [127]
u_b basal velocity [127]
w vertical velocity [160]
x, y, z Cartesian distance coordinates [131]
x_0 initial length [133]
δx incremental length [133]
z distance from glacier bed [128]
δz incremental distance [161]

A_1, A_2 coefficients in Meier's (octahedral) flow law [129]
A coefficient in (octahedral) flow law [160]; area [154]
A' coefficient in (uniaxial compression) flow law [128]
$\bar{A}$ coefficient in (simple shear) flow law [128]
A_0 initial area [154]
B coefficient [128]
$\bar{B}$ coefficient in (effective) flow law [135]
C coefficient [131]; constant of integration [128]
D thickness of flow [128]
E depth of erosion [145]
H heat of fusion of ice [132]
I heat flow per unit time [132]
J volumetric equivalent of heat of fusion of ice [132]
K thermal conductivity of rock obstacle [132]
L size of hypothetical cubical obstacles on glacier bed [130]

L' distance between these cubical obstacles [130]
N difference between normal pressure and water pressure ($= \sigma_1 - p$) [162]
δT pressure melting point difference [131]
V thickness of ice melted in unit time [132]

α surface slope of glacier [135]
β transient creep component [156]
$\epsilon, \dot{\epsilon}$ conventional linear strain (and strain rate) [155]
$\overline{\epsilon}, \dot{\overline{\epsilon}}$ true (or natural) linear strain (and strain rate) [128]
$\epsilon_1, \epsilon_2, \epsilon_3$ principal strains [128]
σ true normal stress [154]
σ_e equivalent stress [157]
σ_r normal pressure exerted by ice on bedrock [158]
$\overline{\sigma}_r, \sigma_r'$ mean and deviate components of this pressure [158]
σ_r'' component of σ_r' acting upvalley [159]
$\Delta\sigma$ maximum value of σ_r' [158]
$\delta\sigma$ incremental normal stress [145]
σ_m mean normal stress [156]
σ_n nominal normal stress [155]
γ bulk unit weight [149]
$\gamma_{eff}, \dot{\gamma}_{eff}$ effective strain and strain rate [156]
$\gamma_{oct}, \dot{\gamma}_{oct}$ octahedral strain and strain rate [129]
ρ, ρ_i mass density of ice [132]
ρ_r bulk mass density of rock [145]
κ Andrade's steady true strain rate [156]
$\overline{\phi}$ mean angle of friction along shear surface [150]
ϕ_i angle of friction at sole-rock contact [150]
τ_0 basal shear stress [130]
τ_{eff} effective shear stress [156]
τ_{oct} octahedral shear stress [129]
λ wavelength of bedrock surface [158]
ω angular frequency ($= 2\pi/\lambda$) [158]

5.1 Simplified glacial motion

Just as the efficacy of fluid erosion depends on the mode of fluid flow, the extent of erosion by moving ice is intimately related to the mechanics of glacier movement. At the outset, it is important to distinguish between cold (or polar) ice masses and temperate ice masses. *Cold* ice is ice in which the temperature is everywhere *below* pressure melting point; in general (Figure 5.1), temperature decreases from the base of a cold ice mass upwards, although there is sometimes an inversion in the uppermost layers. An important feature of cold ice masses is that the ice-rock interface is frozen; meltwater rarely issues from the glacier bed. Furthermore, there is now some evidence to support the idea that the basal velocity of cold glaciers is zero. The erosion potential of such glaciers is thus rather restricted. *Temperate* ice is ice which is *at* pressure melting point throughout; however, it should be recognized that even in so-called temperate glaciers the uppermost layers will be colder than pressure melting point during the winter. The ice–rock interface in temperate glaciers is not frozen and there is assumed to be at least a very thin layer of water between ice and rock. Temperate glaciers are thus capable of sliding over their beds, although evidence (from boreholes and tunnels) indicates that actual velocities vary considerably even within one glacier. Surface velocities u_s are usually greater than basal velocities u_b (Table 5.1); this difference is attributable to shear strain (as described in Section 3.3) developed in response to shear stresses within the ice mass. We shall discuss this internal flow first of all, building on the ideas presented in Chapter 3, and then examine the mechanics of basal sliding.

Application of the ideas relating creep to the internal flow of ice masses is complicated because of the varying cross-sectional geometry of glaciers and ice-sheets. Unless otherwise indicated, we shall for simplicity consider a semi-infinite slab of ice and assume constant slope and ice thickness. Further detail is beyond the scope of this treatment, and the reader is referred to Nye (1965) for a comprehensive discussion of the effect of valley-sides on glacier flow.

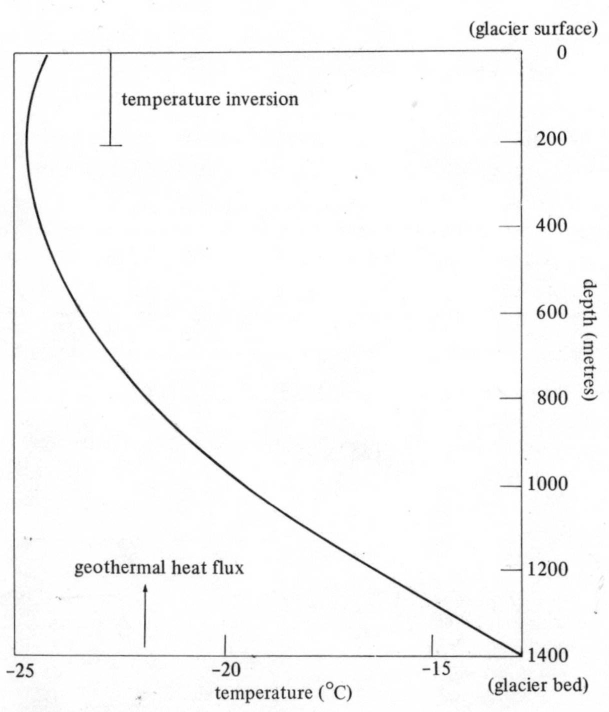

Figure 5.1. Temperature profile in the Greenland ice sheet (Camp Century) (after Hansen and Langway, 1966).

Originally the internal flow of glacial ice downvalley was, by analogy with the flow of water in stream channels, attributed to the deformation of a Newtonian fluid of very high viscosity. Subsequently, uniaxial tests on specimens of polycrystalline ice by Glen (1952, 1955) and others have established without doubt that the viscosity of ice is not independent of the applied stress. Typical results of Glen's (1952) tests were shown in Figure 3.5, and we have already noted (Section 3.3) that, for temperatures between $-0{\cdot}8$°C and $-2{\cdot}5$°C, they may be expressed in the form

$$\dot{\bar{\epsilon}}_1 = A'\sigma_1^n \tag{5.1}$$

with $A' = 0{\cdot}17$ and $n \approx 3$, when $\dot{\bar{\epsilon}}_1$ is in units of years^{-1} and σ_1 is in bars. The exponent n is apparently independent of temperature, but the coefficient A' (a reciprocal measure of viscosity) decreases markedly with lower temperatures. At -13°C, A' is about two orders of magnitude smaller than at 0°C.

Equation (5.1) is widely quoted by glaciologists, but unfortunately in numerous different forms. In order to avoid unnecessary confusion we shall omit these here, but as a guide to following the glaciological literature they are brought together in Appendix 5.1. We may recall from Section 3.3 that a flow law of the form (5.1) is comparable to a velocity profile given by

$$u = -\frac{B}{n+1}(D-z)^n + C\,, \tag{5.2}$$

where $B = 2\bar{A}(\rho g)^n \sin^n i$; $\bar{A}$ is the coefficient in the simple shear flow law analogous to A' for uniaxial compression; and C is the integration

Table 5.1. Measurements of (a) surface velocity and (b) ratio of basal to surface velocity for different glaciers (from Embleton and King, 1968, p.82–83; and Paterson, 1969, p.77).

(a) Glacier	m y^{-1}
South Cascade, USA (Meier and Tangborn, 1965)	20
Saskatchewan, Canada (Meier, 1960)	116
Austerdalsbreen, Norway (in ice fall zone) (Embleton and King, 1968)	2000
Jakobshavn, Greenland (Helland, 1877)	7300
Rinks, Himalayas (Kick, 1957)	10800

(b) Glacier	u_b/u_s
Athabaska, Canada (Savage and Paterson, 1963)	0·10
Mt. Collon, Alps (Haefeli, 1957)	0·20
Salmon, Canada (Mathews, 1959)	0·45
Aletsch, Switzerland (Gerrard, 1952)	0·50
Athabaska, Canada (Savage and Paterson, 1963)	0·75
Blue, USA (Kamb and LaChapelle, 1964)	0·90

constant. Furthermore, it is easily shown [from Equations (3.24) and (3.25)] that the difference between the surface and basal velocities is given by

$$u_s - u_b = \frac{B}{n+1} D^{n+1} \tag{5.3}$$

where D is the thickness of the glacier. (This is identical to Equation 13 of Paterson, 1969, p.92.)

Although this type of velocity profile shows reasonable agreement with a few actually observed profiles, several workers have suggested that Glen's flow law (on which B and n are based) is inaccurate at low stress levels. Meier (1960), for instance, assembled a large amount of laboratory and field data for temperate ice which conformed to a flow law of the type

$$\dot{\gamma}_{oct} = A_1 \tau_{oct} + A_2 \tau_{oct}^{4\cdot5} \tag{5.4}$$

where, for $\dot{\gamma}_{oct}$ in years^{-1} and τ_{oct} in bars, $A_1 = 0{\cdot}018$ and $A_2 = 0{\cdot}13$. Unfortunately, more field observations are really needed before the relative merits of alternative flow laws can be fully assessed. Moreover, there is the further problem of whether the appropriate flow law is affected by the level of hydrostatic pressure. Rigsby (1958) demonstrated that for single crystals there is no effect provided that the temperature is taken relative to pressure melting point rather than 0°C. Paterson (1969) reports that this is also true for polycrystalline ice.

It is probable that in cold ice masses the velocity profile is rather different to that for temperate glaciers, in which, because of isothermal conditions, the same flow law is applicable at all depths. In cold ice masses temperatures are believed to be higher at depth than near the surface; viscosity (as indicated by $1/A'$ in the flow law) should therefore be less, and the strain rate greater, at depth. As a result (Figure 5.2) most of the shear strain should be concentrated in the basal layers of the ice mass, a point emphasized by Nye (1959).

Whereas laboratory studies on the creep of ice have provided reasonable explanations for the internal flow of glacial ice, the issue of basal sliding is still highly controversial. Actually, *prima facie* the remarkable feature of temperate glaciers is not so much that they do slide over their beds, but that they slide so slowly. The resistance to sliding on a *flat* ice–rock interface may be due to two mechanisms [Equation (3.4)]: adhesion between ice and rock, and plane friction along the interface. In temperate glaciers the ice–rock interface is not frozen and therefore the adhesion component is non-existent. Furthermore, it is commonly assumed (e.g. Weertman, 1957, p.34) that there is no friction between the glacier base and the underlying rock along a flat interface. (We shall question this assumption later.) What then is the mechanism preventing temperate glaciers from sliding down-valley at very high velocities? The answer would appear to be the existence of bedrock projections into the glacier base holding back the ice mass; in other words, the ice–rock interface is not completely flat. The important point, however, is that these bedrock projections can only *slow down* the rate of sliding and they cannot eliminate it completely.

The nature of this interrelationship between the geometry of the glacier bed and the rate of basal slip has been examined by several glaciologists, and a heated debate has developed between Weertman in the USA and Lliboutry in France. Weertman (1957) offered two simple mechanisms for basal slip, one based on pressure melting and the other on enhanced creep in the basal layers. Both mechanisms may be described by reference to Figure 5.3a, depicting an idealized glacier bed of cubical protuberances (of size L) separated on a rectangular grid by distances of $L'-L$.

If the basal shear stress acting to drive the glacier forward is denoted by τ_0 (using the notation for boundary shear stress employed in Chapter 2; in terms of the coordinate axes of Figure 3.20, τ refers to τ_{zx}), the total shear force over an area L'^2 is given by $L'^2\tau_0$. Now, this force is being resisted by the normal force from the upvalley side (shaded) of cube B, and, if there is any adhesion between the ice and the downvalley side of cube A, by a tensile force from cube A. [N.B. Whether or not adhesion does commonly occur in the lee of obstacles is a much debated point (see McCall's observations noted on page 140). In the present context, however, this is not an important issue, because the key factor is the pressure *difference* between the two sides of the cube rather than the absolute pressure value at either end.] At equilibrium we therefore have

$$L'^2\tau_0 = \sigma_A L^2 + \sigma_B L^2 \,, \qquad (5.5)$$

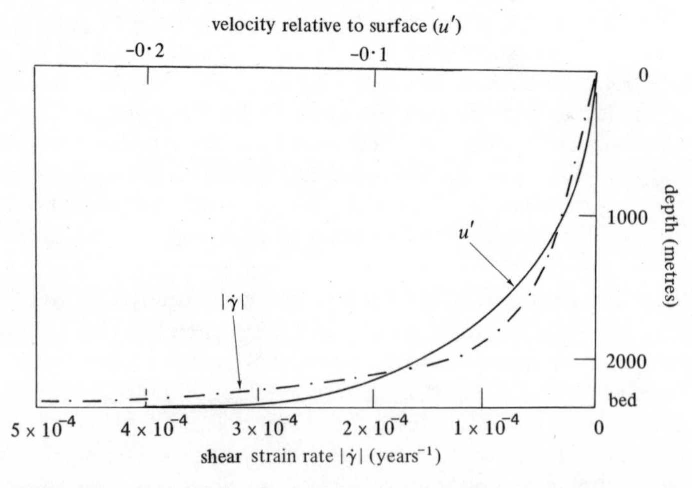

Figure 5.2. Calculated values of velocity and shear strain rate as a function of depth in a cold ice mass (after Nye, 1959).

where σ_A and σ_B are, respectively, the tensile stress from cube A and the compressive stress from cube B on the ice. For a semi-infinite mass with the idealized bed geometry of Figure 5.3a the tensile stresses on the downvalley sides of cubes A and B will be the same. Thus, if we now focus our attention on a single cube (e.g. cube A), the difference in the stress applied to the ice between the upvalley and downvalley sides of the cube is given by

$$\delta p = \sigma_B - (-\sigma_A) = \sigma_B + \sigma_A \,, \tag{5.6}$$

where δp is the pressure (stress) difference. Combining Equations (5.5) and (5.6), we obtain

$$\delta p = \left(\frac{L'}{L}\right)^2 \tau_0 \tag{5.7}$$

as an expression for the pressure difference in the ice on the two sides of the cube. Experiments (cited, for instance, by Smith and Cooper, 1964) have long established that an increase in pressure lowers the melting point of ice at a rate of approximately $7{\cdot}4 \times 10^{-9}$ °C dyne^{-1} cm^2 or 0·0075°C per atmosphere. The excess pressure on the upvalley side of the obstacle must therefore lower the melting point of the ice there by an amount

$$\delta T = C\delta p \,, \tag{5.8}$$

where $C = 7{\cdot}4 \times 10^{-9}$ °C dyne^{-1} cm^2. [Actually, in his 1957 paper, Weertman uses the difference in *mean* pressure $\delta p/3$ in Equation (5.8), but subsequently in his 1964 paper he uses the difference in stress in the direction of flow.] Now, in a temperate glacier the ice is by definition assumed to be at pressure melting point at all places in the glacier. It follows, therefore, that the temperature of the ice just upvalley of an obstacle must be lower than that of the ice just downvalley of it by an

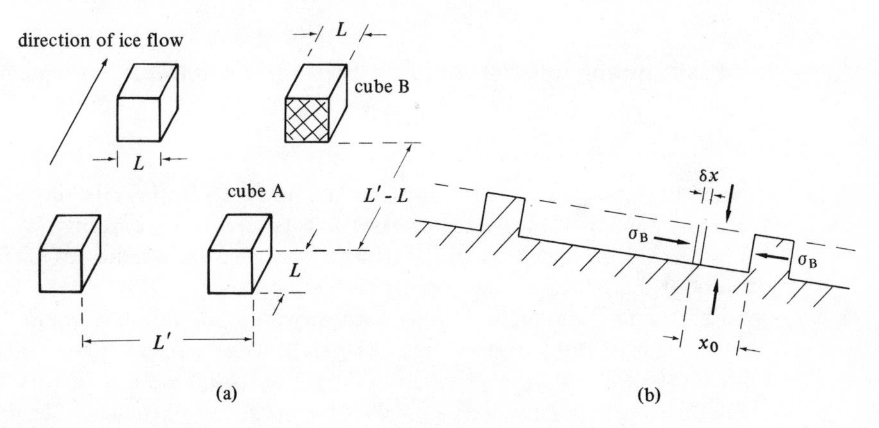

Figure 5.3. Definition diagrams for Weertman's approach to glacier sliding [(a) after Weertman, 1957].

amount δT given by Equation (5.8). In turn, it follows that heat must flow from the downvalley side to the upvalley side; this heat melts ice c the upvalley side and water flows around the obstacle refreezing on the downvalley side. The process is maintained by the release of the latent heat of fusion (80 cal g^{-1}) during refreezing on the downvalley side.

Clearly, this melting of ice on the upvalley side of obstacles, and subsequent flow and refreezing on the downvalley side, is a process of glacier movement. The rate of movement is given by the thickness (in a section along the valley) of ice melted in unit time; quantitatively, it ma be expressed as

$$u_b = \frac{V}{L^2}, \tag{5.}$$

where V is the volume of ice melted upvalley from the obstacle in unit time. Now V is given by the amount of heat flow per unit time upvalley and the amount of heat necessary to melt a unit volume of ice J, that is,

$$V = \frac{I}{J} = \frac{I}{H\rho_i}, \tag{5.1}$$

where J is simply the volumetric equivalent of the heat of fusion of ice H and ρ_i is the mass density of the ice. Moreover, if it is assumed that all the heat flows through the obstacle and none is lost around it, the rate c heat flow is given by

$$I = \frac{\delta T}{L} L^2 K, \tag{5.1}$$

where K is the thermal conductivity of the obstacle. Substituting for V Equation (5.9), we find the rate of basal movement to be

$$u_b = \frac{\delta T L K}{H\rho_i L^2}, \tag{5.1}$$

and substituting for δT [from Equations (5.7) and (5.8)], we obtain

$$u_b = C\left(\frac{L'}{L}\right)^2 \frac{\tau_0 K}{LH\rho_i} \tag{5.1}$$

indicating that, for a given interface geometry L'/L, the rate of basal slip decreases as the typical obstacle size L gets larger. According to Weertman, obstacles of size L = 100 cm would, with values of L'/L equa to about 4, reduce sliding to negligible amounts.

Kamb and LaChapelle (1964) have provided laboratory support for th mechanism of pressure melting (or *regelation*) as a means of ice moveme past obstacles, but Nye (1969) has emphasized the great difficulty in actually applying equations to field situations. In particular, the difficul in estimating an appropriate value for L'/L renders Equation (5.13) of rather restricted practical usefulness.

The second mechanism of basal slip proposed by Weertman (1957, 1964) is essentially accelerated creep in the basal layers. Referring again to Figure 5.3a, the normal stress exerted by the upvalley face of an obstacle on the ice adjacent to it may be expressed [following from Equation (5.5)] by

$$\sigma_B = k\tau_0\left(\frac{L'}{L}\right)^2 \tag{5.14}$$

where the coefficient k depends on the magnitude of the tensile stress σ_A on the downvalley side of the obstacle. If the rock–ice contact on the downvalley face is not frozen, $\sigma_A = 0$ and $k = 1$; if it is assumed that $\sigma_A = \sigma_B$, k is equal to $\frac{1}{2}$. Now, if we consider a small cuboid of ice in contact with the upvalley face of a cube (Figure 5.3b), we observe that it is being compressed by this stress σ_B in a manner which is analogous to, but not identical to, a triaxial compression test. The effect of this compressive stress must be to produce strain in this cube for as long as the stress is applied. It is reasonable to envisage this compression as a squeezing of the ice out of the area upvalley of the obstacle and around it to the downvalley side. By application of Glen's flow law for temperate ice, the strain rate is given by

$$\dot{\bar{\epsilon}}_x = k'A'\left[k\tau_0\left(\frac{L'}{L}\right)^2\right]^n , \tag{5.15}$$

where the coordinate system of Figure 3.20 (x being downvalley) is used and k' is a coefficient which might serve to modify the flow law for uniaxial compression to one that is appropriate for the more complex stress conditions in the problem here. From Chapter 3 we may recall that the strain rate $\dot{\bar{\epsilon}}_x$ is represented by the term

$$\dot{\bar{\epsilon}}_x = \frac{\delta x}{x_0}\frac{1}{\delta t} = \frac{\delta x}{\delta t}\frac{1}{x_0} , \tag{5.16}$$

where δx is the change in length due to compression in the x direction in time δt, and x_0 is the initial length of the specimen under compression in the given stress state. But $\delta x/\delta t$ is, in fact, the rate of compression and in the case under discussion it is the velocity of basal sliding. Substituting for $\dot{\bar{\epsilon}}_x$ in Equation (5.15) we obtain therefore

$$u_b = x_0 k'A'\left[k\tau_0\left(\frac{L'}{L}\right)^2\right]^n . \tag{5.17}$$

Unfortunately, there are many unknowns in this equation. Weertman (1957) *assumes* that $k = \frac{1}{2}$, that Glen's law is directly applicable ($k' = 1$), and that the appropriate distance for x_0 (Figure 5.3b) should be the same as the size of the obstacle L. Equation (5.17) then reduces to

$$u_b = A'\left[\frac{\tau_0}{2}\left(\frac{L'}{L}\right)^2\right]^n L , \tag{5.18}$$

indicating that the rate of basal slip for this mechanism increases as the size of the obstacle. However, it should be emphasized that the transitior from Equation (5.17) to (5.18) is based on unproven assumptions of a questionable nature.

Equations (5.13) and (5.18), derived by Weertman, can be rearranged t yield an expression for the frictional resistance f between the glacier and its bed, per unit area of the interface. Under steady, uniform sliding conditions this resistance must be equal to the driving mechanism of the glacier τ_0, and we obtain an expression of the form:

$$f = cu_b^{1/m} , \qquad (5.19)$$

in which the coefficient c incorporates the interface geometry as well as other components of Equations (5.13) and (5.18) and the exponent is positive (in the pressure melting equation $m = 1$, and in the plastic flow equation $m = n$). Thus, the frictional resistance to sliding provided by the bed of the glacier is itself a function of the velocity of sliding.

Lliboutry (1958, 1965, 1968) accepts in principle the two mechanisms proposed by Weertman, although he uses a different model for the glacier bed (Figure 5.4); working from different assumptions he obtains different equations for the basal velocity. These differences are, however, minor ones. The real contribution made by Lliboutry is his emphasis that water-filled cavities may exist in the lee of protuberances, and that the water pressure in these cavities will exert a significant influence on the rate of basal slip. In his latest discussion of glacier sliding, Lliboutry (1968)

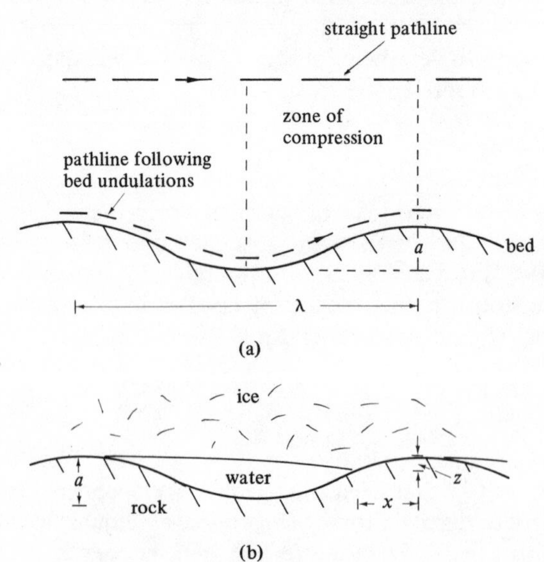

Figure 5.4. Definition diagrams for Lliboutry's approach to glacier sliding (after Lliboutry, 1968).

provides the following equations for the frictional resistance at the glacier bed (assuming sliding due to enhanced plastic creep rather than pressure melting):

$$f = \left(\frac{9}{2}\frac{r^5}{a}\frac{\pi^4}{\overline{B}}u_b\right)^{1/3} \tag{5.20a}$$

and

$$f = \frac{\pi}{6}(\rho g h \cos\alpha - p)^{5/2}\left(\frac{\overline{B}a}{u_b}\right)^{1/2}, \tag{5.20b}$$

the equations relating, respectively, to the cases when cavities do not, and do (with minimal rock–ice contact), exist in the lee of obstacles. The various terms in these equations are indicated in Figure 5.4; the equations are derived in Appendix 5.2.

Equation (5.20a) can be rearranged to yield an expression for the velocity of basal slip which in structure is very similar to Weertman's (5.18), bearing in mind that Lliboutry uses the viscosity coefficient for the effective flow law $\overline{B}$ (see Appendix 5.1.3) and has presented the equation for the particular case of $n = 3$. Equation (5.20b) indicates the key role of the pressure p of the water in the cavities in determining the rate of basal sliding of glaciers. This may be illustrated by considering the effect of a sudden increase in water pressure (as, for instance, might be brought about by an acceleration in the ablation rate) on a glacier sliding at a steady, uniform rate. In accordance with Equation (5.20b) the frictional resistance holding back the glacier would be reduced; initially therefore, the driving mechanism τ_0 would exceed the resistance along the interface and the glacier would accelerate downvalley. Further reference to Equation (5.20b) would seem to indicate continued decrease in the frictional resistance because of the larger u_b value, but other changes must be considered. The increased rate of sliding would in fact produce a reduction in the water pressure (because the cavities would be enlarged) and thus the glacier might be expected to attain a new, but higher, steady velocity of slip. The role of water pressure at the glacier bed is thus very similar to the contribution of excess pore pressures in triggering landslides, as noted in Chapters 3 and 4.

Evidence is accumulating in support of the idea that the presence of water at the ice–rock interface may be an important factor in controlling the velocity of glacier sliding. Elliston (1963) reported that the Gorner Gletscher moves 20–80% more rapidly during summer months than the mean annual velocity; similar observations have been reported by Müller and Iken (1969) for the White Glacier, Axel Heiberg Is., believed to be a cold glacier. Data shown in Figure 5.5 indicate a strong correlation between surface velocity and rate of melt for the White Glacier. Such data do not conclusively indicate that a causal link exists between the abundance of water at the glacier bed and the basal velocity of the glacier, but they do provide some support, in general terms, for Lliboutry speculation. Unfortunately, it is difficult to both predict and measure the water pressure underneath glaciers and at the moment this particular problem must be regarded as the Achilles' heel of this aspect of glaciological research, much as it still is in slope stability studies as noted in Chapter 4

A further point that must be emphasized is the naiveté in the commonly-applied assumption by the glaciologists that on flat surfaces there is no friction along the glacier bed. This assumption completely negates the basic beliefs of geomorphologists interested in glacial erosion. Admittedly, there may be no friction between pure ice and solid bedrock. It is well established, however, that a large proportion of the basal layer of a glacier consists of loose rock and soil particles, in effect bed load; if it is still assumed that plane friction (that is, friction along a flat surface, without interlocking) does not exist at the glacier bed, it becomes rather difficult to explain the occurrence of striae and other evidence of abrasion in glaciated areas.

Finally, it should be appreciated that many glaciers undergo periodic surging at velocities significantly higher than normal (surging glaciers have been known to move as rapidly as 6 km/year for over a year). Various theories have been proposed to explain glacier surges (see Symposium on Surging Glaciers, *Canadian Journal of Earth Sciences,* **6**, No.4, part 2,

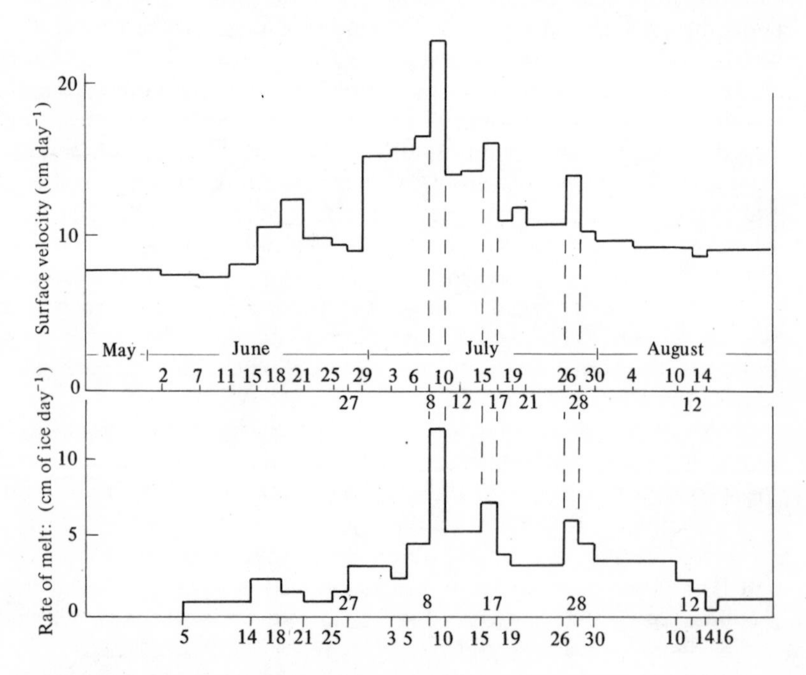

Figure 5.5. Relation between surface velocity and rate of melt for White Glacier, Axel Heiberg Is. (after Müller and Iken, 1969).

1969) and considerable debate still exists. It is known that at high stresses polycrystalline ice is subject to accelerating creep (Section 3.1) and this offers one plausible explanation, although no one has developed this argument in detail. Whatever the explanation for glacier surging, it seems probable that a large amount of glacial erosion may be achieved during these particular events.

5.2 Glacial erosion assuming simplified glacial motion

The efficacy of glacial erosion has been debated on numerous occasions by geomorphologists. Some have held the view that a blanket of glacial ice actually protects the underlying landscape from sub-aerial processes. Others have argued that much of the supposed glacial erosion is in fact essentially fluvio-glacial, attributable to the power of flowing meltwater. For instance, summer meltwater channels are often concentrated on the sides of valley glaciers (Figure 5.6) and the suggestion has been made that the U-shaped valley so typical of glaciated areas may have been produced by accelerated fluvio-glacial erosion along the lateral flanks of valley glaciers. Nonetheless, it is difficult to explain the origin of longitudinal deepening of glaciated valleys, commonly giving rise to lakes and fjords, without resorting to erosion by moving ice. Exactly how glaciers erode the landscape is, however, still relatively unknown. Fluvial geomorphology and related topics have benefited from the application of existing principles from fluid and solid mechanics to these fields (Chapters 2 and 4), but in the study of glacial geomorphology there is still little in the form of a mechanics base which may be utilized. Glaciology is a relatively new discipline and it will probably be some time before the ideas discussed by Paterson (1969) in his *Physics of Glaciers* are applied to the geomorphological problems of the type documented, for instance, by Embleton and King (1968).

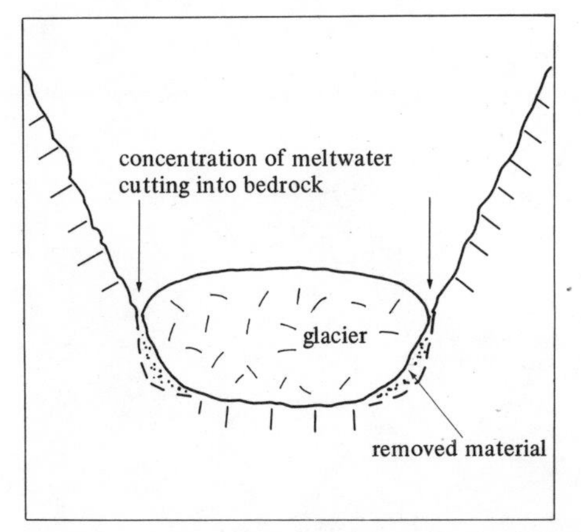

Figure 5.6. Schematic illustration of a possible origin of U-shaped valleys by concentrated meltwater erosion along valley sides.

In fluvial geomorphology a distinction is made between the *wash load* of a stream and the *bed-material discharge.* The former describes sediment which is transported by the stream but derived from outside it, that is, washed into the channel. The latter denotes the transport of debris that is derived from the stream bed. Note that either type of material may be transported as both bed load or suspended load, although for channels lined with coarse debris it is reasonable to assume that bed-material discharge is synonymous with bed load, and wash load with suspended load. Similar terms are appropriate for the transport of material by glaciers. Rock falls and snow avalanches litter the surface of glaciers with debris; this may be regarded as the wash load of the glacier. Much of the material transported by moving ice is also, however, derived from the ice–rock interface, in effect being bed-material discharge, and it is this mode of glacial erosion with which we are concerned here.

Under conditions of simple glacial motion, as discussed previously, it is probable that the bulk of this material is transported in the basal layers of the glacier, analogous to the bed load of streams. McCall (1960) observed such a debris-filled layer (to which he applied the term *sole*, following Garwood and Gregory many years earlier) between the glacier ice and the bed (Figure 5.7) at the head of the Vesl-Skautbreen glacier (Norway). As in the case of streams, this bed load is widely held to be both subject to gradual attrition as it moves downvalley and responsible for corrasion, particularly abrasion, of the underlying bedrock. The production of rockflour and the existence of striae in glaciated areas testify to this abrasion. Although part of the debris in the sole may thus be derived by

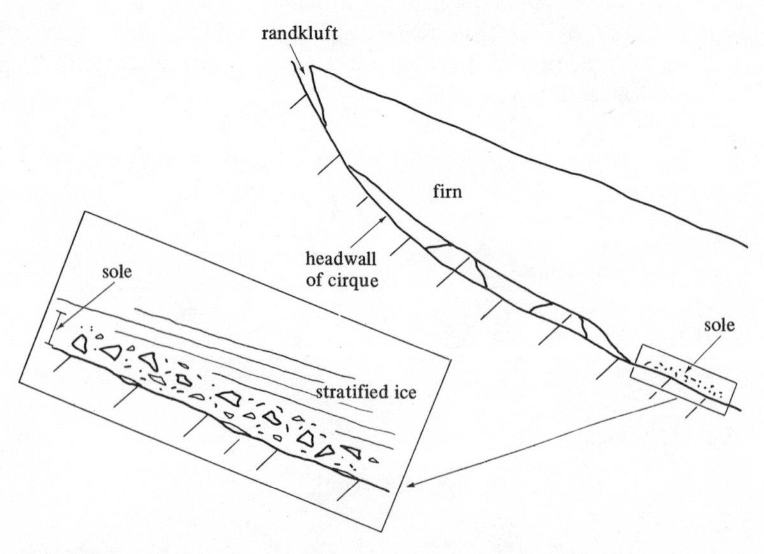

Figure 5.7. Debris in the basal layers of a cirque glacier (Vesl-Skautbreen, Norway) (after McCall, 1960).

further corrasion, the question still remains as to how debris *initially* becomes incorporated into the basal layers of glaciers. One mechanism suggested in the past is plucking of parts of the underlying rock mass by ice, which is frozen to it, as the ice mass moves away. This is now recognized as completely unrealistic, however, because the tensile strength of ice is appreciably lower than the strength of rock, and splitting—if it did occur—would take place either within the ice or at the ice–rock contact rather than within the rock.

It follows that incorporation of rock fragments into the basal layers of an ice mass can take place only if the rock surface underneath the ice is *already* fractured into discrete blocks of rock. This may appear to pose no serious problem, but it must be remembered that the weathered rock mantle of the Earth's surface is, at least in cold areas, relatively shallow (less than 10 m), whereas evidence suggests that glacial erosion has taken place in some cases to depths of over 1000 m. Considerable debate exists as to the mechanism of this glacially-associated weathering, particularly whether it is attributable to deep preglacial weathering or subglacial weathering concurrent with actual glacial transport of debris. This is a point to which we shall return later. At the moment it is necessary to treat the more immediate problem of how loose debris becomes embedded in the basal glacier layers.

Most mechanisms suggested for the incorporation of bed material into the sole of a glacier invoke the freezing of meltwater at the interface between rock and glacial ice. McCall (1960, p.53), for instance, suggested this in the case of the Vesl-Skautbreen:

> "Macroscopically, the ice matrix of the sole was distinctly different from the glacier ice above; it was transparent, bubble-free, layer-free, and full of rock debris ranging in size from fines to boulders. Because of these qualities, such ice must have formed directly from the freezing of water and not from compaction of firn, as does the glacier ice. The sole had a depth of roughly 30 cm and probably originated as follows. Some of the water running down the headwall (of the cirque) finds its way to the bottom of the headwall gap where it freezes. This freezing incorporates the fragments of rock debris lodged there and the whole frozen mass is then pulled along in a continuous manner by the glacier movement, thus forming the sole."

Two mechanisms by which freezing incorporates debris into the basal layers are commonly recognized: one is by plucking and the other by frost heaving.

Various geomorphologists (e.g. Matthes, 1930; Cailleux, 1952) have argued that plucking of *loose* debris is perhaps the most important single process of glacial erosion, but few have examined the process in detail. The regelation mechanism of basal sliding discussed in Section 5.1 does provide, however, theoretical justification for the process. Excess pressure on the upvalley side of obstacles produces melting of ice there and flow to the downvalley side where refreezing occurs. This refreezing may have two effects. Firstly, it could assist in fracturing the downvalley side of the obstacle, and secondly, it could bind discrete fragments of the rock to the moving ice thus leading to plucking. Carol (1947) actually reached a subglacial cavity on the lee side of a roche moutonée beneath the Ober Grindelwald glacier and provided certain evidence to support this view. Kamb and LaChapelle (1964) reached similar conclusions in connection with the Blue Glacier, Washington. One might argue that this form of erosion must be rather restricted because, through this very process, subglacial obstacles would quickly disappear. However, as pointed out by Lewis (1947), at least in the case of roches moutonées, the height of the lee face is often greater than the height of the upvalley side (Figure 5.8), and the process may be a continuing one. Against the concept of glacial plucking, it should be noted that McCall's (1960) study of the base of the Vesl-Skautbreen showed no refreezing in the lee of obstacles there. The problem is, clearly, still open to debate.

The role of frost heaving in the incorporation of debris into the basal glacier layer was suggested by Weertman (1961) for quasi-cold glaciers subject to unsteady flow regimes, based on observations on Baffin Island and near Thule, Greenland. A possible sequence of events is illustrated in Figure 5.9. The top diagram indicates the change in conditions at the glacier base from the edge towards the interior of the ice mass. It is based on the observation that the velocity of glacier movement (as with stream channel flow) is fastest in the interior and decreases towards the edges. Weertman's model assumes, in particular, that this is reflected in the rates of basal sliding and, in turn, this affects the subglacial thermal regime. At the edge of the glacier the bed is assumed to be frozen to the base of the ice, and movement—if there is any—is slow. In the centre of the ice mass the combination of geothermal heat and heat produced from sliding is

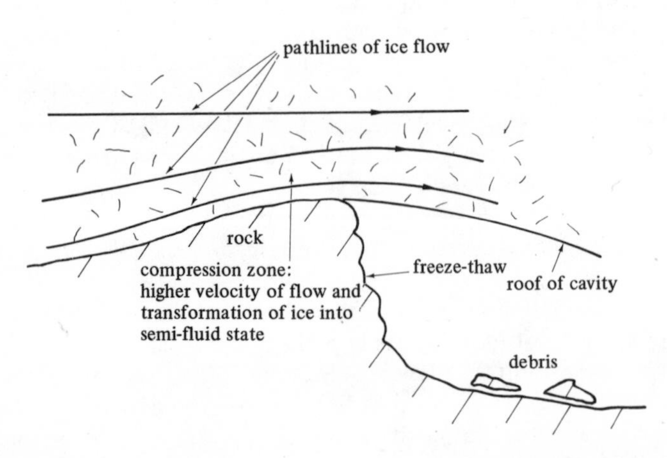

Figure 5.8. Subglacial conditions at a roche moutonnée (after Carol, 1947).

greater than can be conducted away via the temperature gradient in the ice, and as a result ice is melted to water. The pressure in this water forces it to flow towards the flanks of the glacier into regions where the heat input is lower. Accordingly, in this intermediate zone the water refreezes and joins the underside of the ice sheet. The glacier base in this zone is maintained at pressure melting point by the extra latent heat of freezing.

The bottom diagram (Figure 5.9b) illustrates the effect of a temporary reduction in the velocity of the glacier (as might happen through a thinning of the ice mass), followed by a sudden acceleration in the rate of sliding. Initially, the point X (where the 0°C isotherm joins the glacier base) moves towards the interior of the ice mass, thus extending into the glacier the zone in which the rock is frozen to the glacier base. Subsequently, the increased heat input due to the faster rate of basal sliding allows the meltwater from the innermost zone to move further out before completely refreezing. When it moves beyond the point X, it must now flow under the frozen rock debris attached to the glacier base; and when it refreezes, it does so underneath this frozen-on debris. Continuous refreezing and frost heaving is thus believed to push up this layer of

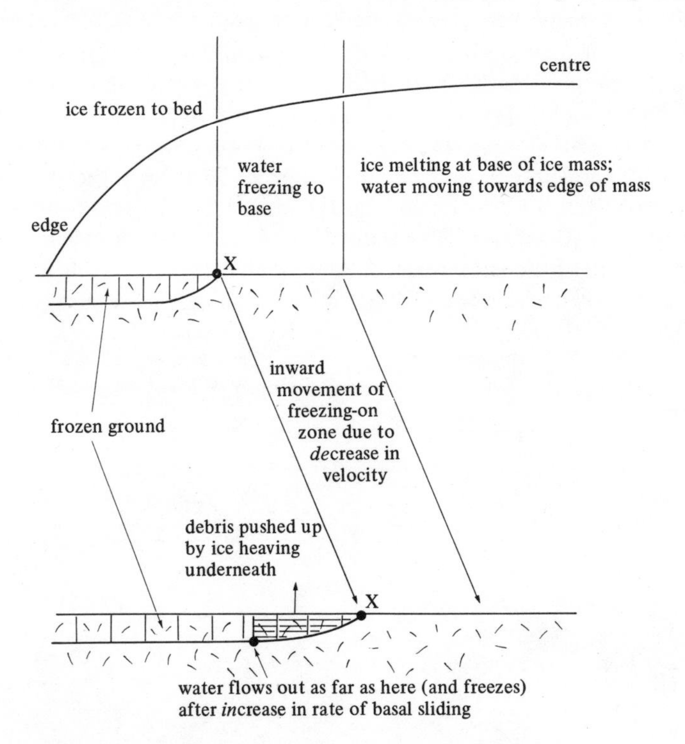

Figure 5.9. Weertman's mechanism of debris incorporation by cold glaciers subject to unsteady flow (based on Weertman, 1961).

frozen rock debris into the ice. Repetitions of this sequence should lead to numerous layers of debris of varying width being incorporated into the ice.

This mechanism is attractive but, even if it is assumed to be valid for quasi-cold glaciers, it is difficult to apply it to temperate glaciers which ar held to be at pressure melting point over the entire glacier base. A simila argument has, however, been presented for the upper areas of Alpine glaciers by Fisher (1953, 1955) who, on the basis of several tunnels dug a far as the glacier bed, maintains that the upvalley parts of Alpine glaciers (usually regarded as temperate ice masses) are even today frozen to the rock floor. The boundary between isothermal ice (where the glacier base is at pressure melting point) and the upper, frozen ice–rock contact is currently at about 4000 m. Fisher argues that oscillation of this contact (up and downvalley), as a result of minor climatic fluctuations, will produce alternate freezing and thawing in this zone leading to shattering o the bedrock. Because the subglacial rock slopes at the heads of Alpine glaciers are so steep, there is little need to invoke actual frost heaving to move debris into the ice but in principle the mechanism is similar to Weertman's, previously described.

As we have already noted, once debris has been incorporated into the basal glacier layers, subsequent erosion is possible through corrasion. Thi may take the form of both abrasion and crushing or planing off of obstacles projecting up from the rock floor. The latter two mechanisms, discussed by McCall (1960), are illustrated in Figure 5.10. Consider the case where a cubical block of rock of volume x^3 (cm^3) is either loaded onto (Figure 5.10a), or abutting against (Figure 5.10b), the upvalley side of a cubical rock projection of volume y^3 (cm^3). For reasons which will become apparent in the subsequent argument, the moving block must be bigger than the obstacle if fracturing of the obstacle is to occur; the

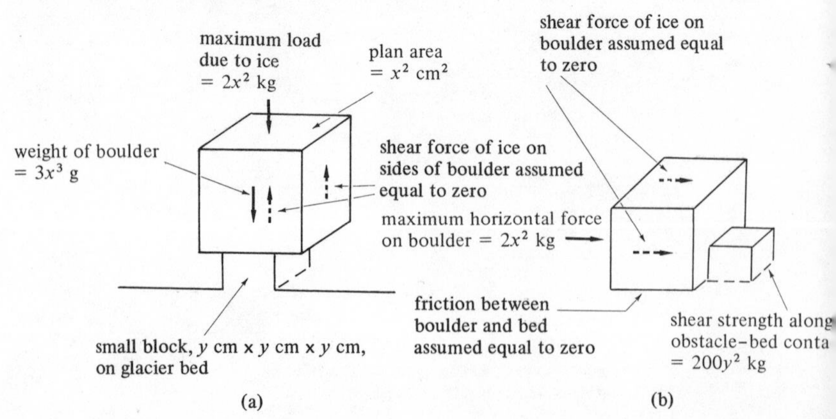

Figure 5.10. Forces involved in (a) crushing and (b) shearing of small obstacle on glacier bed by a loose larger piece of rock.

problem is to calculate the minimum ratio of x/y for dislocation. The crucial parameters in the problem are the strengths of the rock and the ice. The unconfined *compressive* strength of intact specimens of gabbro, granite, limestone, and similar rocks is about 2000 kg cm^{-2}, equivalent to a *shear* strength of about 200 kg cm^{-2}. The unconfined compressive strength of glacier ice is—as noted in Section 3.1—variable, depending on the strain rate. Mellor and Smith (1966) provide a comprehensive set of data on the uniaxial compressive strength of compacted snow. Specimen results are shown in Figure 5.11. At temperatures close to 0°C the strength may reach a maximum of over 10 kg cm^{-2} at strain rates of about $2{\cdot}5 \times 10^{-3}$ s^{-1}; unfortunately, little is known about the actual strain rates in the vicinity of obstacles on the glacier bed, so that it is difficult to choose an appropriate value for the 'strength' of the ice.

McCall (1960) assumed, for illustrative purposes, that ice behaves as a perfectly plastic material with an unconfined *compressive* strength of 2 kg cm^{-2} which, on putting $\sigma_3 = 0$, $\sigma_1 = 2$ kg cm^{-2}, and $b = 45°$ in Equation (1.6), is equivalent to a shear strength of 1 kg cm^{-2}. Adopting these values, we can calculate the minimum x/y ratio for dislocation of the obstacle. In *crushing,* the maximum vertical force that the ice can exert on the detached boulder is $2x^2$ kg; this is supplemented by the weight of the boulder itself on the obstacle and, assuming a unit weight value of 3 gm cm^{-3}, this weight is equal to $0{\cdot}003x^3$ kg. The maximum vertical force on the obstacle is, therefore $(2x^2 + 0{\cdot}003x^3)$ kg. The strength of the obstacle in compression as a force is approximately given by $2000y^2$ kg. At the instant of crushing these two forces are equal and we obtain $x^2(2 + 0{\cdot}003x) = 2000y^2$ or $x/y \approx \sqrt{1000}$, assuming the simplified conditions shown in Figure 5.10a. In lateral *shear* the maximum horizontal force that can be exerted on the boulder by the ice is again $2x^2$ kg (assuming that the sides and top face of the cube are not frozen to the glacier), and in turn this is the maximum force that the boulder can exert on the obstacle. The strength of the obstacle in shear

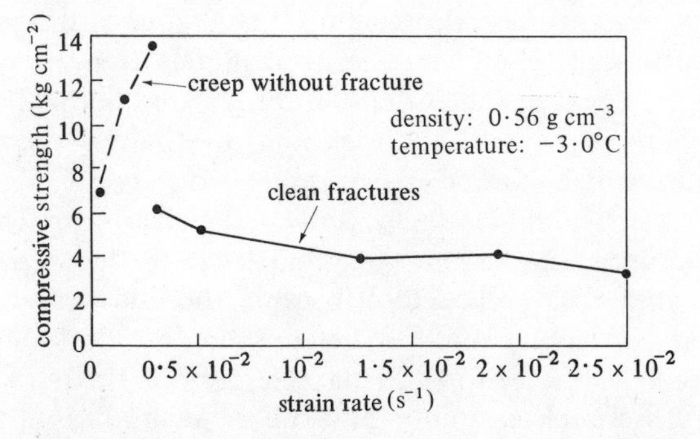

Figure 5.11. Effect of strain rate on strength of snow samples (after Mellor and Smith, 1966).

is, however, only about one-tenth of the compressive strength value. At limiting equilibrium, under the conditions given in Figure 5.10b, we obtain $2x^2 = 200y^2$ or $x = \sqrt{100}y$. In words, in order to shear off obstacles of a given size, a boulder at least ten times this size (in dimensions of length) must be abutting against the obstacle.

The corrasion of the glacier bed by abrasion, crushing, or planing off projections obviously necessitates some prior mechanism for getting loose debris into the basal layers of the ice. Some of these possible mechanisms were discussed previously; all, however, demand that the rock immediately underneath the glacier is already fractured. Some workers believe that subglacial conditions themselves favour weathering of the bedrock concurrent with glacial erosion; others argue that this is unlikely and that deep glacial erosion can only occur if there is deep preglacial weathering of the bedrock. Some of these points are touched upon below.

Workers who attribute glacial erosion to freeze–thaw processes, such as plucking of loose debris and heaving of debris into the ice, have little difficulty in explaining the preparatory weathering of the bedrock in order to produce this debris. Alternate freezing and thawing should be just as effective in opening up and extending fractures in the rock as in the actual process of incorporating it into the basal ice. However, it is very difficult to apply this generally to temperate glaciers which, by definition, are assumed to be at pressure melting point; meltwater will not refreeze simply because it is in contact with ice, if the ice is at pressure melting point. The classic temperature measurements made by Battle and Lewis (1951) in bergschrunds in Greenland have emphasized that, although marked temperature fluctuations may occur in the low atmosphere, temperatures at the ice–rock interface of many glaciers show only small changes from 0°C.

Because of this problem, other possible mechanisms of subglacial weathering have been sought. Prominent among these ideas is the suggestion made by Lewis (1954), and others, that erosion itself is a mechanism of weathering in the form of pressure-release on the bedrock surface. This concept was introduced previously in Chapter 4, where it was noted that the cutting of valleys in cohesive material often produces tension cracks in the valley side. These tension cracks are due to the release of *lateral* pressure at points in the rock mass by the removal of the rock (by valley cutting) which exerted that pressure. Inevitably, similar effects must occur in the rock under valley floors; removal of material during the cutting and deepening of the valley must reduce the vertical pressure at any point under the valley floor, because the thickness of the overburden is reduced. This relaxation of the confining pressure must produce some expansion of the bedrock in the direction of the reduced stress. As a result, pressure-release joints (also called *dilatation* joints) roughly parallel to the ground surface are developed. Simultaneous cracking in other directions will assist in fracturing the bedrock into discrete fragments. If a comparison is made between the pressure at a given depth under a valley floor before glaciation and at the same point after 100 m of glacial erosion, an appreciable reduction in the vertical pressure will have taken place at that point. The actual amount, of course, depends on the thickness of glacial ice. The relaxation in vertical pressure is approximately given by

$$\delta\sigma = \rho_r gE - \rho_i gD \quad (5.21)$$

where $\rho_r g$ is the unit *bulk* weight of the rock removed, $\rho_i g$ is the unit weight of the ice, E is the depth of erosion, and D is the ice thickness. If we assume for illustrative purposes that E = 100 m, D = 50 m, $\rho_r g$ = 2300 kg m^{-3} and $\rho_i g$ = 900 kg m^{-3}, then the relaxation in vertical pressure would be approximately 185 000 kg m^{-2} (260 psi). The actual amount of strain produced by this pressure release would depend on the modulus of elasticity of the rock complex.

This argument may well seem irrelevant in the initial stages of glaciation (prior to any glacial erosion) but it should be remembered that, at least in the case of valley glaciers, substantial pressure release and concomitant dilatation cracking may have been already produced by preglacial stream downcutting. In addition, several other mechanisms of preglacial fracturing of the valley bedrock have been suggested by other workers. Boyé (1950), for instance, argued that extensive freezing of groundwater in valley bottom areas prior to glaciation would produce appreciable break-up of the bedrock through normal frost shattering; and Cailleux (1952) and Harland (1957) have added variations to this theme.

5.3 Erosion under more complex modes of glacier flow

In the simplified glacial motion described in Section 5.1 and assumed in Section 5.2 ice moves in bands parallel to the main slope of the glacier bed. Minor protuberances produce localized departures from this average condition, but these local contortions in the pathlines of the flow are restricted to the basal layer. The beds of actual glaciers inevitably, however, contain undulations of a much larger scale than these minor bumps, and these must distort the flow pattern of the ice body. As shown in Figure 5.12, large-scale undulations in the flow of ice take place reflecting the geometry of the underlying valley long-profile. If it is assumed that glacier discharge is constant over a reach of valley (between tributary glaciers), changes in bedrock slope must produce changes in glacier velocity and thickness. Where the gradient increases downglacier, velocity (which has been shown to be a function of basal shear stress and, therefore, a function of the valley gradient) will increase, and (if discharge is constant) the thickness of flow must decrease. Conversely, ice thickness will increase and velocity decrease in areas where the gradient decreases downglacier. Nye (1952) referred to these two types of flow as *extending* (thinning) and *compressive* (thickening), and pointed out that they correspond to the *active* and *passive* Rankine states of stress used in soil mechanics. This may be appreciated by referring back to Section 3.2; active failure corresponds to conditions produced by stretching (downslope) a soil mass, and passive denotes compression of the soil mass as, for instance, by pushing a retaining wall upslope. This observation by Nye has important implications for glacier flow and erosion. Reference to Figure 3.17 shows that in neither case are the shear surfaces parallel to the surface of the ice mass (as in the flow described in Section 5.1) and, moreover, that there is a marked difference in the orientation of the shear surfaces in the two cases. These differences are shown more clearly in Figure 5.13.

In the extending (active) case the shear surfaces along which flow occur are oriented into the bedrock surface so that ice flows (obliquely) towards the ground. In the compressive (passive) case ice flows obliquely away from the bed of the glacier. Moreover, in the active failure of an ideal rigid plastic material a tension zone develops and tension cracks open. Although ice is not perfectly plastic, analogous cracks, crevasses, occur in the active flow of glaciers. Note that the flow patterns shown in Figure 5.13 do assume that the ice is isotropic; zones of weakness in the ice would modify this picture.

The implications of this contrast between extending and compressive flow can be appreciated by reference to Figure 5.14 combining the two modes of flow over a basin-and-step profile. In the upper parts of the basin (zone A) flow is toward the bedrock; similarly, movement of englacial debris is towards the rock surface. Although it is possible that such action increases the abrasive action in this zone, it seems more probable that, in fact, *net* removal of debris from the ground in this zone will be reduced. A situation in which debris is moving from within the glacier to the basal layers is hardly a conducive environment for

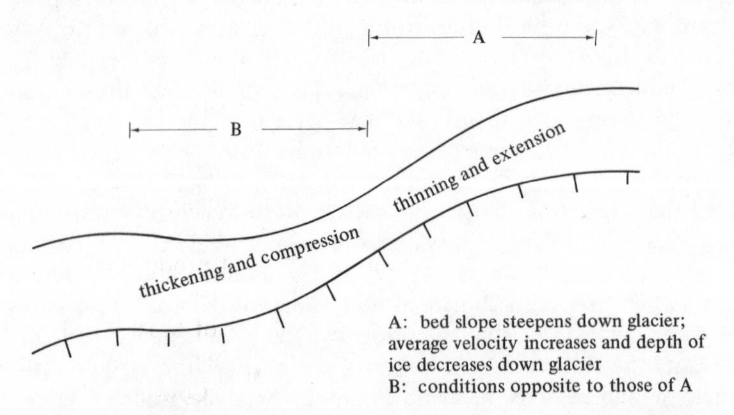

Figure 5.12. Influence of large scale bed topography on mode of glacier flow.

incorporating extra debris into the basal layers from the rock surface beneath. In contrast, in the lower part of the basin (zone B) the continuous movement of ice and 'bed load' away from the glacier bed might be expected to increase the rate of net movement of debris out of this zone and increased removal of material should take place. Superimposed on these differences due to the contrast in flow patterns is the effect of crevasse development. Crevasses are restricted to fairly thin zones (Appendix 5.3) and as a result the probability of crevasses reaching down to bedrock is inversely related to the thickness of the glacier. It follows that the most likely location at which crevasses will extend down to the glacier bed is at the junction between zones A and B. This, as

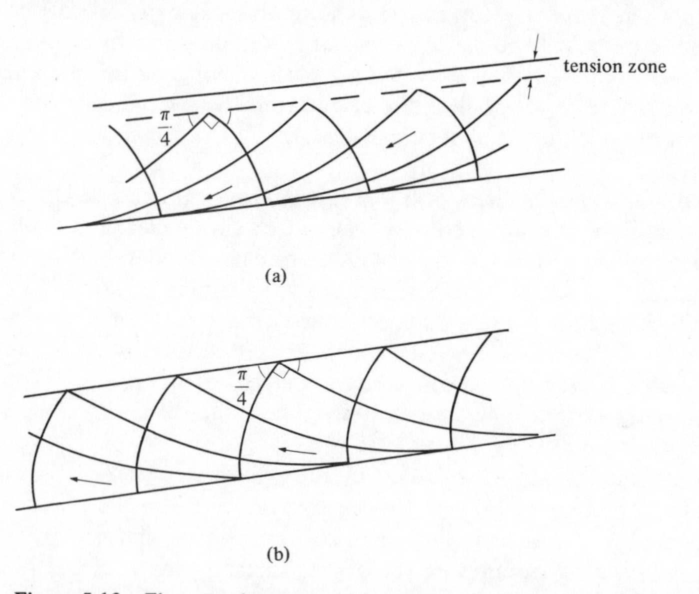

Figure 5.13. Flow conditions in (a) extending (active) case and (b) compressive (passive) case (refer back to Figure 3.17).

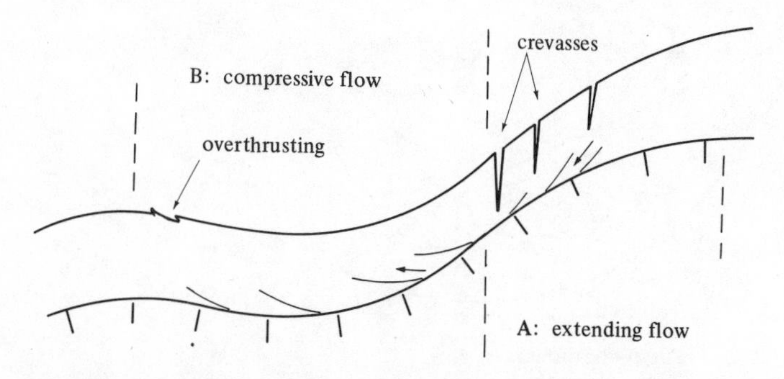

Figure 5.14. Schematic illustration of flow conditions over undulating bed.

emphasized by Lewis (1947), is important, because it means that cold air and meltwater have access to the glacier bed at this point, and as a result this location should be a favourable one for enhanced freeze–thaw action In combination, these various features of the ice flow over an undulating surface suggest that, at a macro-scale, glacial erosion should accentuate th preglacial relief, rather than subdue it, and this may explain the basin-and step profiles characteristic of many glaciated valleys.

Inasmuch as a *cirque* can be regarded as merely the uppermost basin o a glacial stairway, it is possible that the preceding discussion has some bearing on the formation of this landform too. Most geomorphologists, however, tend to attribute much of the excavation of cirques to glacial periods when glaciers occupied only the uppermost parts of valleys; that is, cirques are usually regarded as primarily the products of erosion by cirque glaciers rather than the upvalley parts of long glaciers extending much further downvalley than the cirque rim. Some evidence for this is the frequent occurrence of moraine dumped at the downvalley edge of cirques.

Studies of cirque glaciers over the last twenty-five years (e.g. Lewis, 1960) suggest that much of the movement of cirque glaciers is *rotational*, analogous to the rotational earth slips mentioned in Section 4.4. The chief difference is that, whereas an earth *slip* involves the rotation of a rigid soil mass over a single curved failure surface, rotational *flow* in cirq glaciers involves some internal flow as well. Observations by McCall (1960) have, however, revealed that the movement of the Vesl-Skautbree approximates very closely to a rigid-body motion, with over 90% of the surface movement being due to basal sliding.

One of the first suggestions that cirque glacier movement was rotation was made by Lewis (1947) on the basis of ice layers in these glaciers (Figure 5.15). These layers are separated by mineral and organic materia which, where they outcrop on the glacier surface, may accumulate as

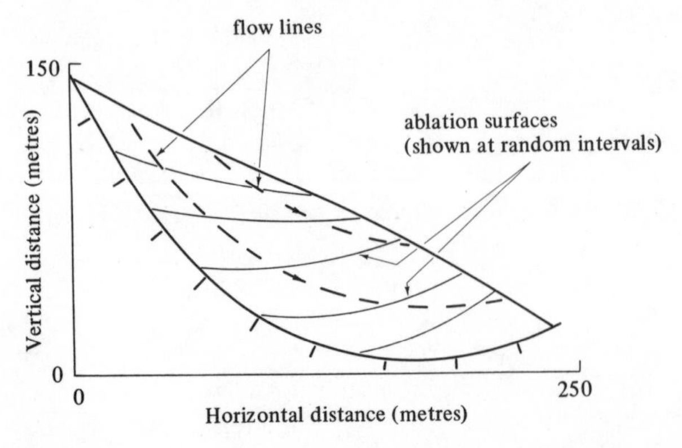

Figure 5.15. Long section through Vesl-Skautbreen glacier (after McCall, 1960).

surface debris. Lewis speculated that these 'dirt bands' between the ice layers might be rotational shear surfaces and that the surface debris might have been derived from corrasion of the cirque floor. Subsequent work showed that this interpretation was invalid. Each ice layer in fact represents one year's accumulation of snow in the firn region of the glacier, and the 'dirt bands' separating them represent ablation and accumulation of some debris during intervening summer months. This annual accumulation of snow in the firn zone acts as the driving force behind the slow rotation of the glacier and, as additional layers of firn accumulate each year, old layers are moved down through the glacier, giving rise to the appearance shown in Figure 5.15. On the Vesl-Skautbreen the ice layers are about 1 m thick which, assuming the unit weight of ice to be 0·9 gm cm^{-3}, represents about 15 m of fresh snow at a unit weight of about 0·06 gm cm^{-3}. Not all of this accumulation represents *in situ* snowfall and some of it has undoubtedly been derived from drifting. This may offer an explanation why so many of the most spectacular cirques occur on the lee side of mountains: if the rate of cirque erosion is dependent on the rate of rotation of the glacier, and this in turn reflects the rate of accumulation in the firn zone, erosion should be most intense where drifting accumulates most snow.

Clark and Lewis (1951) used this rotational motion of cirque glaciers in explaining the origin of the cirque landform. They argued that abrasion at the rock–ice interface under the rotating ice mass must mould the bedrock surface into an arcuate form producing the typical long profile of cirques. *Prima facie*, there may appear to be a chicken–egg element in this argument: do glaciers occupying cirques rotate because the bedrock surface already has an arcuate form, or is the basin profile a genuine effect of rotational glacial erosion? Clark and Lewis (1951) favour the latter, and their approach may be supported by application of conventional slope-stability analysis to cirque glaciers. Figure 5.16a shows the probable preglacial profile of a valley head with a distinct break of slope at the junction of the valley floor and the end-slope. Consider any potential failure arc as shown in the diagram. From Chapter 4 (Figure 4.9) we know that for a $\phi = 0$ material (as ice is commonly regarded) the stability number varies between about 7·5 and 3·8, depending on the angle of the slope surface. In relation to Figure 5.16a this means that, assuming for illustration that $i = 15°$, if the stability number exceeds $7c/\gamma$, a rotational failure will occur along the indicated arc, irrespective of the geometry of the bedrock surface underneath. Substituting $c = 1$ kg cm^{-2} (the commonly accepted nominal yield stress for glacial ice) and $\gamma = 0{\cdot}9$ gm cm^{-3}, the critical H value is equal to 63 m. This, of course, does not mean that at smaller H values there is no rotational movement (because ice is not an ideal rigid plastic material and has no true yield stress), but rather at larger values the rotational velocity increases rapidly. Lewis (1960) used this mode of approach in his argument that a tendency to rotate exists in almost all cirque glaciers. Unfortunately the problem is not quite this simple, because, if abrasion is taking place at the ice–rock interface, it is invalid to assume that at the interface $\phi = 0$.

In the situation depicted in Figure 5.16a only a small portion of the potential failure surface is actually part of the ice–bedrock interface and therefore the average frictional element over the slip surface would be very small. The type of analysis suggested above may therefore be quite realistic. If rotational movement is assumed to take place on this slip surface, erosion should be concentrated in the zones of contact between the bedrock and the shear surface, as indicated in the diagram. Continue erosion over a period of time would thus mould the bedrock into an arcuate form (Figure 5.16b), even though the initial subglacial profile was not curved. Indeed, the more rounded the bedrock profile becomes (as the glacier cuts deeper) the less effective erosion should be, because the driving force of the rotation is being resisted by an increasing amount of friction as the shear surface–bedrock contact area increases. The average friction $\overline{\phi}$ on the shear plane can be expressed *approximately* as

$$\tan\overline{\phi} = a\tan\phi_i \text{ ,} \qquad (5.2$$

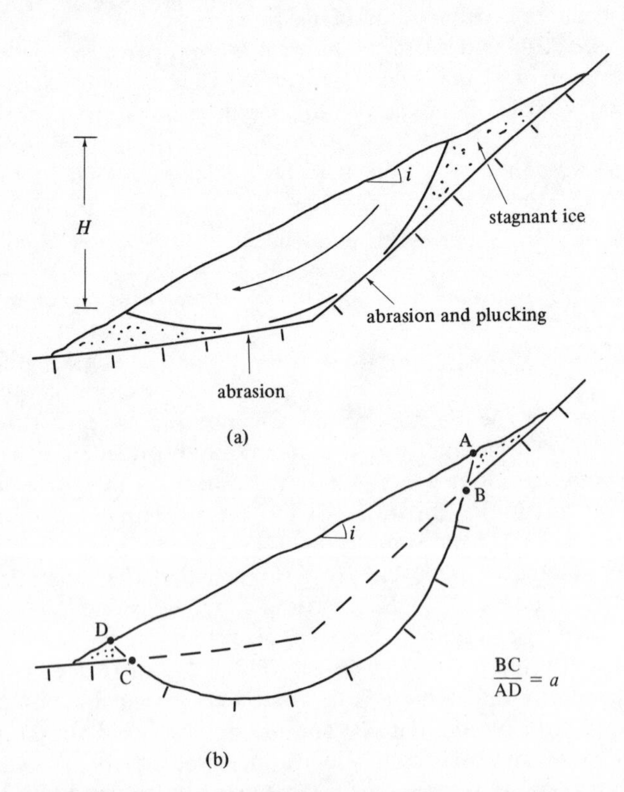

Figure 5.16. Rotational flow theory of cirque erosion.

where a is the fraction of the shear surface in contact with the bedrock and ϕ_i is the friction of the bedrock contact. For $a = 0$, $\bar{\phi} = 0$, and for $a = 1$, $\bar{\phi} = \phi_i$. Presumably, ϕ_i is approximately 35°, the angle of friction of loose rock rubble ($\tan\phi_i = 0 \cdot 71$), and reference to Figure 4.9 shows that, if this is so, the erosional ability of the glacier would decrease quite rapidly as it cut into the rock mass. As an example, for $\bar{\phi} = 25°$ ($a = 65\%$), H would need to be at least $20c/\gamma$—that is, 200 m, even with a 40° slope—in order to drive an abrading glacier, assuming a nominal yield stress of 1 kg cm^{-2} for glacial ice.

The rotational theory of cirque glacier erosion is an attractive one but, because of the difficulty of determining the frictional resistance at the rock surface under the glacier, it is difficult to evaluate it fairly. Moreover, a large body of geomorphologists still believe that the essential process of cirque erosion is backwearing of the headwall (rather than over-deepening of the valley floor) due to frost shattering. In this case the mechanics of glacial erosion would be a rather irrelevant aspect of the problem, the key element being weathering. White's (1970) recent comments, based on the examination of cirque form in the Rockies, provide, however, a certain amount of support for the over-deepening hypothesis of cirque formation.

5.4 End-note

The application of mechanics to glacial erosion is a frustrating task for several reasons. First, the discipline of glaciology is relatively new, and it is only recently that basic aspects of glacial mechanics have begun to emerge. The tremendous dependence on the data of Glen's pioneer experiments, contrasted to the wealth of existing data in soil mechanics available for slope stability studies, emphasizes the need for more basic information about the mechanical properties of glacial ice. Second, there is the immense difficulty of gaining access to the ice–rock interface of deep glaciers. Drilling is beginning to yield some information, but it is a much more difficult task than coring into clay masses. This is particularly so on cold glaciers. Third, it may well emerge that the central problem in glacial erosion is not so much the mechanics of subglacial transport of debris but the preparation of debris available for movement at the subglacial ground surface. Even less is known about this aspect of the problem. It is possible, as recognized by some workers, that both rock breakdown and transport are due to glacier movement, but the actual answer will only emerge when much more is known about both conditions at the glacier bed and the mechanics of glacier ice.

Bibliography

Battle, W. R. B., Lewis, W. V., 1951, "Temperature observations in bergschrunds and their relationship to cirque erosion", *J. Geol.,* **59**, 537-545.

Boyé, M., 1960, *Glaciaire et Périglaciaire de l'Ata Sund Nord-Oriental (Groenland)* (Expéditions Polaires Françaises, 1, Paris).

Cailleux, A., 1952, "Polissage et surcreusement glaciaires dans l'hypothèse de Boyé", *Revue Géomorph. Dyn.,* **3**, 247-257.

Carol, H., 1947, "Formation of roches moutonées", *J. Glaciol.,* **1**, 57-59.

Clark, J. M., Lewis, W. V., 1951, "Rotational movement in cirque and valley glaciers", *J. Geol.,* **59**, 546-566.

Elliston, G. R., 1963, Discussion of a paper by J. Weertman, *Bull. Int. Ass. Scient. Hydrol.,* **8**, 65-66.

Embleton, C., King, C. A. M., 1968, *Glacial and Periglacial Geomorphology* (Edward Arnold, London).

Fisher, J., 1953, "Two tunnels in cold ice at 4000 m on the Breithorn", *J. Glaciol.,* **2**, 513-520.

Fisher, J., 1955, "Internal temperatures of a cold glacier and conclusions therefrom", *J. Glaciol.,* **2**, 583-591.

Glen, J. W., 1952, "Experiments on the deformation of ice", *J. Glaciol.,* **2**, 111-114.

Glen, J. W., 1955, "The creep of polycrystalline ice", *Proc. Roy. Soc. (London), Ser.A,* **228**, 519-538.

Hansen, B. L., Langway, C. C., 1966, "Deep-core drilling and ice core analysis, Camp Century, Greenland, 1961-1966", *Antarctic Journal of the United States,* **1**, 207-208.

Harland, W. B., 1957, "Exfoliation joints and ice action", *J. Glaciol.,* **3**, 8-10.

Kamb, W. B., LaChapelle, E. R., 1964, "Direct observation of the mechanism of glacier sliding over bedrock", *J. Glaciol.,* **5**, 159-172.

Lewis, W. V., 1947, "Valley steps and glacial valley erosion", *Trans. Inst. Br. Geogr.,* **14**, 19-44.

Lewis, W. V., 1954, "Pressure release and glacial erosion", *J. Glaciol.,* **2**, 417-422.

Lewis, W. V., 1960, "The problem of cirque erosion", in *Norwegian Cirque Glaciers,* Royal Geographical Society Research Paper No.4, pp.97-100.

Lliboutry, L., 1958, "Contribution à la théorie du frottement du glacier sur son lit", *Compt. Rend. Hebd. Séanc. Acad. Sci.,* **247**, 318-320.

Lliboutry, L., 1965, *Traité de Glaciologie*, Tome 2 (Masson, Paris).

Lliboutry, L., 1968, "General theory of subglacial cavitation and sliding of temperate glaciers", *J. Glaciol.,* **7**, 21-58.

Lliboutry, L., 1969, "The dynamics of temperate glaciers from the detailed viewpoint", *J. Glaciol.,* **8**, 185-205.

McCall, J. G., 1960, "The flow characteristics of a cirque glacier and their effects on glacial structure and cirque formation", in *Norwegian Cirque Glaciers,* Royal Geographical Society Research Paper No.4, pp.34-62.

Matthes, F. E., 1930, "Geologic history of the Yosemite valley", *U. S. Geol. Surv. Profess. Paper* 160.

Meier, M. F., 1960, "Mode of flow of the Saskatchewan Glacier, Alberta, Canada", *U. S. Geol. Surv. Profess. Paper* 351.

Mellor, M., Smith, J. H., 1966, "Strength studies of snow", Cold Regions Research and Engineering Laboratory Research Report 168, Hanover, New Hampshire, USA.

Mendelson, A., 1968, *Plasticity: Theory and Application* (MacMillan, New York).

Müller, F., Iken, A., 1969, "Velocity fluctuations and water region of arctic valley glaciers", Paper presented at Symposium on the Hydrology of Glaciers (Cambridge, September 1969), to be published by International Association of Scientific Hydrology.

Nye, J. F., 1952, "The mechanics of glacier flow", *J. Glaciol.,* **2**, 82-93.
Nye, J. F., 1957, "The distribution of stress and velocity in glaciers and ice sheets", *Proc. Roy. Soc. (London) Ser.A,* **239**, 113-133.
Nye, J. F., 1959, "The motion of ice sheets and glaciers", *J. Glaciol.,* **3**, 493-507.
Nye, J. F., 1965, "The flow of a glacier in a channel of rectangular, elliptic or parabolic cross-section", *J. Glaciol.,* **5**, 661-690.
Nye, J. F., 1969, Discussion of a paper by L. Lliboutry, *Canadian Journal of Earth Sciences,* **6**, 952.
Paterson, W. S. B., 1969, *The Physics of Glaciers* (Pergamon Press, Oxford).
Rigsby, G. P., 1958, "Effect of hydrostatic pressure on velocity of shear deformation of single ice crystals", *J. Glaciol.,* **3**, 273-278.
Smith, A. W., Cooper, J. N., 1964, *Elements of Physics* (McGraw-Hill, New York), p.271.
Weertman, J., 1957, "On the sliding of glaciers", *J. Glaciol.,* **3**, 33-38.
Weertman, J., 1961, "Mechanism for the formation of inner moraines found near the edge of cold ice caps and ice sheets", *J. Glaciol.,* **3**, 965-978.
Weertman, J., 1964, "The theory of glacier sliding", *J. Glaciol.,* **5**, 287-303.
Weertman, J., 1967, "An examination of the Lliboutry theory of glacier sliding", *J. Glaciol.,* **6**, 489-494.
White, W. A., 1970, "Erosion of cirques", *J. Geol.,* **78**, 123-126.

Appendix 5

5.1 Glen's stress–strain rate law for polycrystalline ice

The experimental data on which this 'law' is based were derived from a series of uniaxial compression tests. In order to understand fully these results, three points must be emphasized: (1) Glen's data are reported in terms of *natural* strain rather than *conventional* strain; (2) Glen provides two separate sets of data, one including the *transient* creep component and the other attempting to eliminate it; and (3) subsequent workers (Nye, Meier, Lliboutry, and others) have expressed Glen's original $\dot{\bar{\epsilon}}_1$–σ_1 data in the form of a more general flow law, each using a different set of terms. These three points are discussed below.

5.1.1 *Natural and conventional strain*

Much of the previous discussion in this book has implicitly treated stress–strain relations in terms of *nominal stress* and *conventional strain* rather than *true stress* and *natural* (or *true*, or *logarithmic*) *strain*. Previously there was little need to mention this difference. Consider a uniaxial compression test on a cylindrical specimen of initial area A_0 and length l_0, subjected to a load P. The initial compressive stress is:

$$\sigma_{\text{initial}} = \frac{P}{A_0}, \tag{A5.1}$$

as we have pointed out before. During the test, the specimen is being compressed and therefore the length is being reduced and the area enlarged (Figure A5.1). If the substance is incompressible (as ice is usually considered to be),

$$Al = A_0 l_0 . \tag{A5.2}$$

At any time after the test has started, the *true* stress is given by

$$\sigma = \frac{P}{A} \tag{A5.3}$$

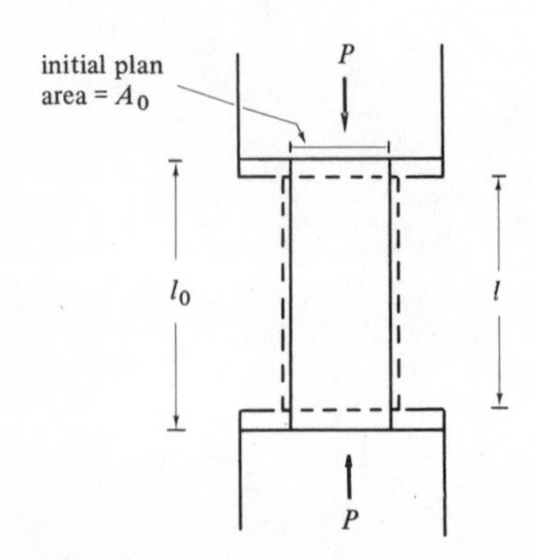

Figure A5.1. Definition diagram for true stress and strain compared to nominal stress and conventional strain during axial compression.

and the stress P/A_0 is only a *nominal* stress (often denoted by σ_n). We thus have:

$$\frac{\sigma}{\sigma_n} = \frac{A_0}{A} \; . \tag{A5.4}$$

Now, we have already defined the *conventional* linear strain as:

$$\epsilon = \frac{l-l_0}{l_0} = \frac{l}{l_0} - 1 \tag{A5.5}$$

and this can be written [from (A5.2)] as

$$\epsilon = \frac{A_0}{A} - 1 \; . \tag{A5.6}$$

From Equations (A5.4) and (A5.6) it follows that the true stress is given by

$$\sigma = \sigma_n(1+\epsilon) \tag{A5.7}$$

so that for infinitesimal strain the true stress is approximately given by the nominal stress.

Similarly, the *conventional* linear strain relates to the amount of deformation relative to the initial length. If a specimen of initial length l_0 is strained to length l_1 and then an additional compression to length l_2 is performed, the true strain is $(l_2 - l_1)/l_1$ rather than $(l_2 - l_0)/l_0$, which is the conventional strain. Ludwik therefore defined the *true* strain in going from length l_0 to l as

$$\bar{\epsilon} = \int_{l_0}^{l} \frac{\mathrm{d}l}{l} = \ln\frac{l}{l_0} \; . \tag{A5.8}$$

The true strain is related to the conventional strain quite simply; from Equations (A5.5) and (A5.8), we have

$$\bar{\epsilon} = \ln(1+\epsilon) \; . \tag{A5.9}$$

This distinction between natural (true) and conventional strain is important at large strains. The condition of incompressibility for a substance is usually defined (using principal strains) by

$$\epsilon_1 + \epsilon_2 + \epsilon_3 = 0 \; ,$$

but this is only true for natural strain values, although it may be taken as valid for *small* conventional strains.

5.1.2 *Transient and steady creep*

Typical stress–strain curves are shown in Figure 3.4. The general curve has four components: (i) an immediate initial elastic strain, (ii) transient strain (attenuating creep), (iii) steady creep, and (iv) accelerating creep leading to progressive failure. The strain rate used in Glen's papers refers to the *minimum* strain rate measured in a test; assuming that the period of transient creep is over, this describes the steady creep rate. However, it seemed probable that some of the tests undertaken by Glen had not reached the steady creep phase when they were stopped, so that, particularly at low stresses, some of the $\dot{\bar{\epsilon}}$ values may be misleading. Some technique for separating the transient and steady components of the creep curve was thus necessary. Glen employed Andrade's well-known law:

$$\frac{l}{l_0} = (1+\beta t^{1/3})\mathrm{e}^{\kappa t} , \tag{A5.10}$$

where t refers to the elapsed time and β indicates the transient component. Using Equation (A5.8), we can rephrase this:

$$\bar{\epsilon} = \ln(1+\beta t^{1/3}) + \kappa t \tag{A5.11}$$

which for $\beta = 0$ would reduce to

$$\bar{\epsilon} = \kappa t , \tag{A5.12}$$

or, differentiating with respect to t,

$$\dot{\bar{\epsilon}} = \kappa ; \tag{A5.13}$$

κ is thus the steady true strain rate. By fitting Equation (A5.11) to his data, Glen was able to obtain an estimate of κ for each test. His results (for tests conducted at 0°C) gave the relationship

$$\kappa = \dot{\bar{\epsilon}}_1 = 0{\cdot}017\sigma_1^{4\cdot 2} . \tag{A5.14}$$

The main batch of Glen's data was, however, not corrected by this method, but, because values are available for different temperatures, these other results are more widely quoted. The typical expression is

$$\dot{\bar{\epsilon}}_1 = A'\sigma_1^n , \tag{5.1}$$

where $n = 3{\cdot}17$ (and is relatively unaffected by temperature), and A' has the values 0·17 (0°C), 0·023 (−1·5°C), 0·008 (−6·7°C), and 0·0017 (−13°C) at different temperatures (these values of A' and n are only valid when σ_1 is expressed in bars and $\dot{\bar{\epsilon}}_1$ in years^{-1}).

5.1.3 *Alternative forms of the flow law*

As noted previously, Meier (1960) cites Glen's data in terms of an octahedral shear stress and shear strain rate. This poses no problem because these terms have been fully discussed in Chapters 1 and 3. Nye (1957, 1959), in contrast, re-expresses Glen's data in terms of an *effective shear stress* (we shall denote this by τ_{eff}, for clarity, although Nye simply uses τ) and an *effective strain rate* ($\dot{\gamma}_{\mathrm{eff}}$, rather than Nye's $\dot{\epsilon}$). These parameters are defined as follows:

$$\tau_{\mathrm{eff}}^2 = \tfrac{1}{2}[(\sigma_x-\sigma_m)^2+(\sigma_y-\sigma_m)^2+(\sigma_z-\sigma_m)^2+2(\tau_{xy}^2+\tau_{yz}^2+\tau_{zx}^2)] , \tag{A5.15}$$

$$\dot{\gamma}_{\mathrm{eff}}^2 = \tfrac{1}{2}[\dot{\epsilon}_x^2+\dot{\epsilon}_y^2+\dot{\epsilon}_z^2+2(\dot{\gamma}_{xy}^2+\dot{\gamma}_{yz}^2+\dot{\gamma}_{zx}^2)] , \tag{A5.16}$$

which can be shown to reduce to

$$\tau_{\text{eff}} = \sqrt{\frac{3}{2}}\tau_{\text{oct}} , \tag{A5.17}$$

$$\dot{\gamma}_{\text{eff}} = \sqrt{\frac{3}{2}}\dot{\gamma}_{\text{oct}} . \tag{A5.18}$$

(This reduction is quite straightforward but rather lengthy and for this reason it is not given here.) Using Equations (A5.17) and (A5.18) [and recalling Equations (3.11) and (3.16)], we find that *in the uniaxial case* Nye's parameters reduce to

$$\tau_{\text{eff}} = \sqrt{\frac{3}{2}} \times \frac{\sqrt{2}}{3}\sigma_1 = \frac{\sigma_1}{\sqrt{3}} , \tag{A5.19}$$

$$\dot{\gamma}_{\text{eff}} = \sqrt{\frac{3}{2}} \times \frac{1}{\sqrt{2}}\dot{\epsilon}_1 = \frac{\sqrt{3}}{2}\dot{\epsilon}_1 . \tag{A5.20}$$

Glen's data (A5.14) may thus be expressed in terms of Nye's parameters as

$$\frac{2}{\sqrt{3}}\dot{\gamma}_{\text{eff}} = 0{\cdot}017\,(\sqrt{3}\tau_{\text{eff}})^{4\cdot 2}$$

or

$$\dot{\gamma}_{\text{eff}} = 0{\cdot}148\tau_{\text{eff}}^{4\cdot 2} \tag{A5.21}$$

which is the equation used by Nye (1957, p.129).

More important, and the real advantage in using these parameters, is that for simple shear [recalling Equations (3.12) and (3.18)] we have

$$\tau_{\text{eff}} = \sqrt{\frac{3}{2}} \times \sqrt{\frac{2}{3}}\tau_{zx} = \tau_{zx} , \tag{A5.22}$$

$$\gamma_{\text{eff}} = \sqrt{\frac{3}{2}} \times \sqrt{\frac{2}{3}}\dot{\gamma}_{zx} = \dot{\gamma}_{zx} , \tag{A5.23}$$

and accordingly Glen's flow law (A5.14) reduces in the case of simple shear to

$$\dot{\gamma}_{zx} = 0{\cdot}148\tau_{zx}^{4\cdot 2} , \tag{A5.24}$$

comparable to (A5.21). [Unfortunately, Nye's *effective shear* stress is easily confused with the *effective* (or *equivalent*) stress ($\sigma_e = 3\tau_{\text{oct}}/\sqrt{2} = \sqrt{3}\tau_{\text{eff}}$) a commonly-used parameter in theoretical mechanics (Mendelson, 1968, p.102). Moreover, note that neither of these parameters is at all connected with the principle of effective stress, as used in soil and rock mechanics, described in Chapter 4!]

The casual reader of the glaciological literature may be confused further by parameters introduced by Lliboutry (1969) who uses the same terms as Nye (effective shear stress and effective strain rate), but although the former is the same in both cases, the effective strain rate as defined by Lliboutry (denoted simply by $\dot{\gamma}$) is twice that employed by Nye. Presumably this reflects the two definitions of shear strain used in mechanics, noted previously in Chapter 3 (p.84). For this reason, Lliboutry's statement of Glen's flow law (A5.14) takes the form:

$$\dot{\gamma}_{\text{eff}} = 0{\cdot}296\tau_{\text{eff}}^{4{\cdot}2}\,, \tag{A5.25}$$

compared to Equation (A5.21) from Nye.

5.2 Development of Lliboutry's equations for basal slip (in relation to the basal shear stress) resulting from plastic creep (regelation ignored)

5.2.1 *No cavitation case (Figure A5.2)*

The profile of the glacier bed, in the simplest case, is assumed to be given by

$$z = \tfrac{1}{2}a\cos\omega x \tag{A5.26}$$

where $\omega = 2\pi/\lambda$. (More realistic profiles are provided by superimposing several periodic curves, of varying wavelength and amplitude, on each other.) If the normal pressure that the ice exerts on the bedrock is defined by

$$\sigma_{\text{r}} = \bar{\sigma}_{\text{r}} + \sigma_{\text{r}}' \tag{A5.27}$$

as in the diagram, it follows from (A5.26) that

$$\sigma_{\text{r}}' = -\Delta\sigma\sin\omega x \tag{A5.28}$$

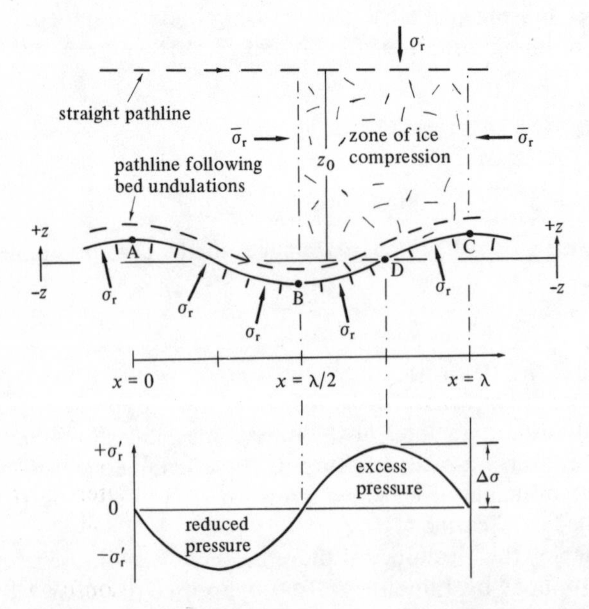

Figure A5.2. Definition diagram for Lliboutry's approach to glacier sliding due to enhanced plastic creep *without* cavitation (after Lliboutry, 1968).

where σ_r' is the fluctuation of this pressure relative to the mean value $\overline{\sigma}_r'$, and $\Delta\sigma$ is the *maximum* deviation.

Consider the forces in the average downglacier direction AC over a distance λ. The driving force is λf (where f is Lliboutry's notation for the basal shear stress). The resistance is provided by the component of the stress exerted by the bedrock on the ice σ_r acting against the direction of motion, integrated over the distance λ. Between A and B this component is negative; the bedrock between A and B is in effect adding to the downglacier force on the ice. Between B and C this component is positive. Over the distances AB and BC the forces due to the mean stress $\overline{\sigma}_r$ are equal in magnitude ($\frac{1}{2}\lambda\overline{\sigma}_r$), but opposite in sign, and therefore they can be ignored. The resistance to glacier motion over the length λ (which at equilibrium is equal to the driving force) is therefore given by

$$f\lambda = \lambda \int_0^\lambda \sigma_r'' \, dz \, , \qquad \text{(A5.29)}$$

where σ_r'' is the component of σ_r' in the average direction upslope CA. The relation between σ_r' and σ_r'' is, of course, given by the triangle of stresses, the difference between σ_r' and σ_r'' being attributable to the shear stress along that part of the bed:

$$\sigma_r''^{\,2} + \tau^2 = \sigma_r'^{\,2} \, . \qquad \text{(A5.30)}$$

For temperate glaciers, it is usual to assume that $\tau = 0$, and thus $\sigma_r'' = \sigma_r' = \Delta\sigma \sin \omega x$. Equation (A5.29) thus becomes

$$f\lambda = -\lambda \int_0^\lambda \Delta\sigma \sin \omega x \, dz \, , \qquad \text{(A5.31)}$$

or, after obtaining the derivative dz/dx from (A5.26),

$$f\lambda = \frac{a\omega\lambda\Delta\sigma}{2} \int_0^\lambda \sin^2 \omega x \, dx \qquad \text{(A5.32)}$$

$$= \frac{a\omega\lambda\Delta\sigma}{2} \left[-\frac{\sin \omega x \cos \omega x}{2\omega} + \frac{x}{2} \right]_0^\lambda$$

$$= \frac{a\omega\lambda\Delta\sigma}{4}$$

or, substituting $\omega = 2\pi/\lambda$,

$$f = \frac{a\pi\Delta\sigma}{2\lambda} \, . \qquad \text{(A5.33)}$$

Now, $\Delta\sigma$ can be shown to relate to the basal velocity and, if this relationship is substituted into Equation (A5.33), an equation linking u_b and f results. Lliboutry has offered a number of possible approaches to this problem and the treatment below follows his 1968 paper. Recall Glen's flow law (5.1). From the development of Equation (A5.21) it can be seen that this law may be expressed in terms of Nye's effective parameters as

$$\dot{\gamma}_{\text{eff}} = \frac{(\sqrt{3})^{n+1}}{2} A'(\tau_{\text{eff}})^n \tag{A5.34}$$

and in terms of Lliboutry's effective parameters (taking $n = 3$) as

$$\dot{\gamma}_{\text{eff}} = \bar{B}\tau_{\text{eff}}^3 \tag{A5.35}$$

where $\bar{B}$ (the coefficient for the effective flow law of Lliboutry that is analogous to A for the octahedral flow law) is equal to $9A'$. Recall also Equations (A5.17) and (A5.18), and (1.11) and (3.14). We shall now substitute for the principal stresses in terms of the state of stress shown in Figure A5.2. Because the shear stress over any small part of the glacier bed is zero, σ_r must be a principal stress; Lliboutry denotes it σ_3. The other principal stresses are given by

$$\sigma_1 = \sigma_2 = \bar{\sigma}_3 \neq \sigma_3 \,. \tag{A5.36}$$

Using this information we find the effective shear stress

$$\tau_{\text{eff}} = \sqrt{\frac{3}{2}\frac{\sqrt{2}}{3}}\,|\,\sigma_1 - \sigma_3\,| = \frac{1}{\sqrt{3}}|\,\bar{\sigma}_3 - \sigma_3\,| = \frac{1}{\sqrt{3}}\sigma_3' = \frac{1}{\sqrt{3}}\sigma_r' \,. \tag{A5.37}$$

Under these stress conditions (A5.36) (bearing in mind the incompressibility condition, assumed for ice, introduced in Appendix 5.1.1) the corresponding principal strain rates are

$$\dot{\epsilon}_1 = \dot{\epsilon}_2 = -\tfrac{1}{2}\dot{\epsilon}_3 \,. \tag{A5.38}$$

On using Equations (A5.18) and (3.14), the effective strain rate then reduces to

$$\dot{\gamma}_{\text{eff(L)}} = 2\dot{\gamma}_{\text{eff(N)}} = 2\sqrt{\frac{3}{2}}\frac{\dot{\epsilon}_3}{\sqrt{2}} = \sqrt{3}\dot{\epsilon}_3 \,, \tag{A5.39}$$

where the subscripts L and N refer to Lliboutry and Nye. Substituting for these parameters in (A5.35) we obtain

$$\sqrt{3}\dot{\epsilon}_3 = \bar{B}\left(\frac{\sigma_r'}{\sqrt{3}}\right)^3$$

or

$$\dot{\epsilon}_3 = \frac{\bar{B}}{9}(\sigma_r')^3 \,. \tag{A5.40}$$

Let us now consider the strain rate at point D where $\sigma_r' = \Delta\sigma$. The downglacier velocity (parallel to x axis) *at* D is denoted by u_b, and the velocity parallel to the z axis is indicated by w. The relationship between these two velocities is given by

$$\frac{w}{u_b} = \frac{dz}{dx} , \tag{A5.41}$$

where dz/dx is the slope of the glacier bed. Differentiation of (A5.26) indicates that at D (where $x = \frac{3}{4}\lambda$),

$$\frac{dz}{dx} = \tfrac{1}{2}a\omega = \frac{\pi a}{\lambda} \tag{A5.42}$$

and, therefore,

$$u_b = \frac{w\lambda}{\pi a} ; \tag{A5.43}$$

u_b is claimed by Lliboutry to be an *approximate* value for the *average* basal velocity over the bed.

Finally, consider the strain in the z direction at D; this is given by

$$\dot{\epsilon}_{z_0} = \frac{\delta z}{z_0}\frac{1}{\delta t} = \frac{w}{z_0} . \tag{A5.44}$$

Lliboutry assumes that $z_0 = \frac{1}{4}\lambda$ and $\dot{\epsilon}_3 = \dot{\epsilon}_{z_0}$, and obtains a new expression for the basal velocity [from (A5.43)]:

$$u_b = \dot{\epsilon}_3\frac{\lambda}{4}\frac{\lambda}{\pi a} = \frac{\dot{\epsilon}_3\lambda^2}{4\pi a} \tag{A5.45}$$

or, rearranging,

$$\dot{\epsilon}_3 = \frac{4\pi a u_b}{\lambda^2} . \tag{A5.46}$$

Substituting for $\dot{\epsilon}_3$ in Equation (A5.40), we obtain

$$\frac{4\pi a u_b}{\lambda^2} = \frac{\overline{B}}{9}\sigma_r'^{\,3}$$

and [from (A5.33)] with $\sigma_r' = \Delta\sigma$,

$$\frac{4\pi a u_b}{\lambda^2} = \frac{\overline{B}}{9}\left(\frac{2\lambda f}{a\pi}\right)^3$$

or

$$f = \left(\frac{9\pi^4 a^4 u_b}{2\lambda^5 \overline{B}}\right)^{1/3} \tag{A5.47}$$

or, defining a bed roughness $r = a/\lambda$,

$$f = \left(\frac{9}{2}\frac{r^5}{a}\frac{\pi^4 u_b}{\overline{B}}\right)^{1/3}$$

which is Lliboutry's (1968) equation 7.

5.2.2 *Case with cavitation (Figure A5.3)*
The situation is more complicated in this case. In relation to Figure A5.3, Lliboutry computes (through a simple but lengthy procedure) that the average basal shear stress is given by

$$f = \tfrac{1}{2}\pi r N \frac{(\pi s - \frac{1}{2}\sin 2\pi s)\sin(\pi s - \omega x_c)}{\sin \pi s - \pi s \cos \pi s}, \tag{A5.48}$$

where $N = \rho g h \cos\alpha - p = \sigma_1 - p$, and p is the pressure of the water in the lee of obstacles. The 'friction' value f thus varies according to the degree of cavitation s. For illustrative purposes, we shall consider here only the case where s is very small, which Lliboutry shows to correspond to $\omega x_c \approx \pi s^2$. In this case (A5.48) becomes:

$$\begin{aligned} f &\approx \tfrac{1}{2}\pi r N \times 2\pi s(1-s) \\ &\approx \tfrac{1}{2}\pi r N \times 2\pi s \,. \end{aligned} \tag{A5.49}$$

The problem now is to relate the degree of cavitation s to the rate of basal sliding u_b. Between A and Y there is a downward movement of the glacier base at a velocity denoted by w. Recalling Equation (A5.44), we

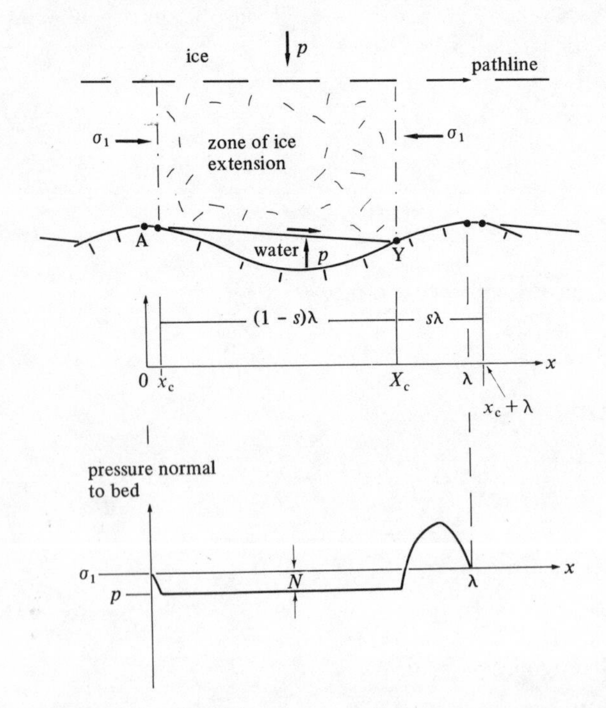

Figure A5.3. Definition diagram for Lliboutry's approach to glacier sliding due to enhanced plastic creep *with* cavitation (after Lliboutry, 1968).

can write the strain rate involved in this process as

$$\dot{\epsilon}_3 = \frac{4w}{\lambda} \; . \tag{A5.50}$$

This strain process is closely analogous to that in a uniaxial compression test under a stress N and [from (A5.40)] with σ_r' being comparable to N we obtain

$$\dot{\epsilon}_3 = \frac{4w}{\lambda} = \frac{\bar{B}}{9}N^3$$

or

$$w = \frac{\bar{B}N^3\lambda}{36} \; . \tag{A5.51}$$

Now, w/u_b is an expression for the slope of the roof of the cavity; at the point where cavitation begins (x_c, z_c) this is the same as the slope of the bedrock surface which [by differentiating (A5.26)] is equal to $\pi r \sin \omega x_c$. Substituting for w in (A5.51), we obtain

$$\pi r \sin(\omega x_c) = \frac{\bar{B}\lambda N^3}{36 u_b} \tag{A5.52}$$

and (for large cavities), substituting $\omega x_c \approx \pi s^2$,

$$s \approx \left(\frac{\bar{B}\lambda N^3}{36\pi^2 r u_b}\right)^{1/2} \; . \tag{A5.53}$$

If we now substitute for s in Equation (A5.49), we can derive an equation linking the basal velocity and the average basal shear stress:

$$f = \frac{\pi}{6}N^{5/2}\left(\frac{\bar{B}a}{u_b}\right)^{1/2}$$

where, to repeat, $N = (\sigma_1 - p)$.

5.3 Prediction of the depth of crevasses in areas of extending flow

5.3.1 *Ice assumed to be perfectly plastic with a yield strength in compression of 2 kg cm^{-2}*

Application of Equation (4.7) [if we bear in mind that $2c\tan(45° + \phi/2) = q_u$ as shown in Appendix 4.6] yields

$$z_0 = \frac{q_u}{\gamma} = \frac{2000}{0{\cdot}9} \text{ cm} \approx 20 \text{ m}$$

for the thickness of the tension zone. This is the maximum depth of crevasses. The reasoning behind this approach is quite simple and is illustrated in Figure A5.4. For values of h such that $\gamma h > q_u$, failures will occur closing up the gaps between the blocks.

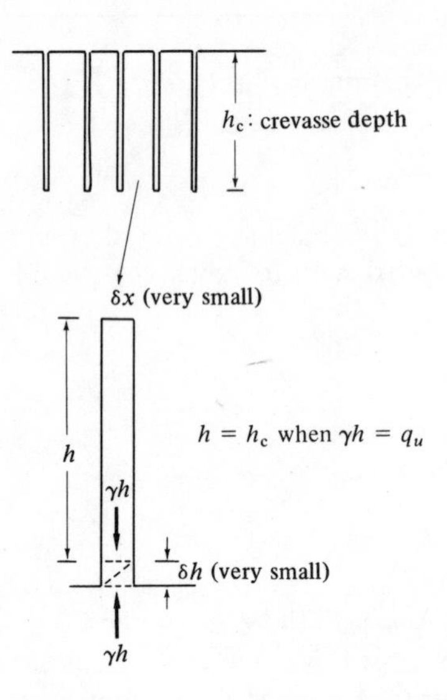

Figure A5.4. Interpretation of maximum depth of crevasses assuming perfect plasticity.

5.3.2 *Prediction based on Glen's flow law*

Referring again to Figure A5.4, we note that the ice at the base of a block is being strained vertically at a rate $\dot{\epsilon}_z$. For a two-dimensional situation, and on assuming ice is incompressible, all of this vertical compression is translated into a horizontal flow at a rate $\dot{\epsilon}_x = \dot{\epsilon}_z$. For any one block half of this flow enters both of the adjacent crevasses. For any one crevasse the gap is being closed at a rate of $\frac{1}{2}\dot{\epsilon}_x$ from both sides, that is at a rate of $\dot{\epsilon}_x = \dot{\epsilon}_z$. Now, the vertical stress at depth z is given by

$$\sigma_z = z(\rho_i g)$$

and substituting this into Glen's flow law (5.1), we obtain

$$\dot{\epsilon}_z = A'[z(\rho_i g)]^n \tag{A5.54}$$

as an expression for the rate of closing of the crevasse.

At the same time the crevasses are being opened up at a rate equal to the longitudinal strain rate of the glacier which may be defined as

$$\dot{\epsilon}_x = \frac{u_A - u_B}{L}$$

where u_A and u_B are velocities (downglacier) at two points separated by a distance L. (It is, of course, precisely such a difference in velocity that produces the stretching of the ice which is responsible for the crevasses.)

At equilibrium, the rate of crevasse opening is just balanced at depth z by the rate of closing of the crevasses. This depth (the limit of crevasse penetration) is thus given by

$$\dot{\epsilon}_x = A'(z\rho_i g)^n$$

or

$$z = \frac{(\dot{\epsilon}_x/A')^{1/n}}{\rho_i g} . \qquad \text{(A5.55)}$$

There is *no* unique value for crevasse depth. It depends on the temperature of the ice (A' becoming smaller, and thus z becoming larger, for colder ice), and also it depends on the rate of stretching $\dot{\epsilon}_x$. Paterson (1969) cites $\dot{\epsilon}_x = 0{\cdot}5$ year^{-1} as being a typical value for this rate; for temperate ice (taking $n = 3$ and $A' = 0{\cdot}17$) Equation (A5.55) yields in this case

$$z \approx 15 \text{ m}$$

which is of the same order of magnitude as the value derived by treating ice as a perfectly plastic material.

Conclusion

The neglect of the study of geomorphic processes by workers in geomorphology must rank as one of the most puzzling features in the development of this discipline. There are those who put the blame for this on the writings of W. M. Davis, arguing that his all-embracing model of landscape evolution left little for the geomorphologist to do besides applying the model successively to new areas. And yet this is hardly fair. Two decades ago, Strahler tried valiantly to infuse a genetic approach into the subject, but was greeted with meagre response. To be sure, the challenge of his quantitative research was met by a surge of field studies, but, sadly, little was undertaken in the way of providing a rational framework for the collection of such data, as his *Dynamic Basis of Geomorphology* had urged. Even today, geomorphologists seem to show a marked reluctance to delve into the genetic, as against the descriptive, side of landscape studies. There is something strange about a subject in which its research workers are willing to dabble at the application of the jargon of thermodynamics, but unwilling to apply even the most rudimentary aspects of mechanics to these problems. One suspects that geomorphology will emerge as a reputable discipline only when its students have become well-versed in the established principles of natural science.

The application of mechanics to geomorphological studies has already offered solutions to many hitherto unanswered problems. In Chapter 2 we observed that the classic concave long-profile of streams may be explained quite simply in terms of the mechanics of debris movement at a fluid boundary. It was also noted that the concept of grade is capable of field testing, provided that field data are examined in relation to the mechanics of fluid flow. Moreover, such studies are not restricted to stream channels, but could be applied to the effects of overland flow on hillslope development. In Chapter 4 it was shown that the height and steepness of cliffs and hillslopes can be explained in terms of the mechanical properties of the materials into which they are cut. And, by observing how these properties change under weathering, it is possible to construct slope development sequences on a rather more precise basis than those offered by earlier workers. Although comparable achievements are much fewer in glacial geomorphology, it is nevertheless clear that only through the application of similar physical principles will problems such as the origin of cirques, the development of fjords, and others be solved satisfactorily. It is to be hoped that future developments in glacial geomorphology and glaciology will not occur in isolation from each other.

At the same time, the study of the mechanics of erosional processes is only one aspect of the study of landscape development. Throughout this essay reference has been made to the importance of understanding the nature of weathering processes preparatory to actual erosion. This was discussed in relation to glacial erosion; similar problems occur in connection with fluvial erosion. Studies of stream channel erosion (as described in Chapter 2) have tended to focus on the mechanics of debris transport, while relatively little attention has been given to the preparation of this debris by subfluvial weathering. Many, perhaps most, erosional processes are to varying degrees weathering-limited and on a geological timescale the mechanics of erosion must be closely linked with the mechanics and chemistry of rock breakdown.

One last point needs to be made. In the last decade remarkable specialization has occurred within the field of geomorphology, just as within the field of Earth science generally. It would be unfortunate if this specialization were allowed to become incorporated into the teaching of the subject. It cannot be overstressed that the underlying principles of landscape development are remarkably similar in all the subdisciplines of the subject, whether we are dealing with coastal, fluvial, glacial, or some other area of geomorphology. One of the purposes of this monograph has been to draw together the various strands of specialized research in the field of process geomorphology in the hope that the student of the subject will appreciate its underlying unity, rather than be alarmed by its superficial diversity.

Bibliography

Bagnold, R. A., 1953, *The Physics of Blown Sand and Desert Dunes* (Methuen, London).

Carson, M. A., Kirkby, M. J., 1971, *Hillslope Form and Process* (Cambridge University Press, Cambridge).

Leopold, L. B., Wolman, M. G., Miller, J. P., 1964, *Fluvial Processes in Geomorphology* (Freeman, San Francisco).

Scheidegger, A. E., 1970, *Theoretical Geomorphology* (Springer-Verlag, Berlin).

Strahler, A. N., 1952, "The dynamic basis of geomorphology", *Bull. Geol. Soc. Am.*, **63**, 923-938.

Yatsu, E., 1966, *Rock Control in Geomorphology* (Sozosha, Tokyo, Japan).

Sample problems

1.1 In an area of flat ground the water table is at the ground surface and the water body is virtually stagnant. What is the pore water pressure (the pressure of water in the pores) at a depth of 3 m? What is the total normal stress due to the weight of the soil acting on a plane parallel to the ground surface at a depth of 3 m? Assume that the unit weight of the water body and the saturated unit weight of the soil mass are 1000 kg m^{-3} and 2200 kg m^{-3} respectively.

(Answer: 3000 kg m^{-2}; 6600 kg m^{-2})

1.2 A cubic block weighs 60 kg in air and 40 kg when immersed in water. Compute (a) the volume of the block and (b) its specific gravity, assuming that the unit weight of the water is 1000 kg m^{-3}.

(Answer: 0·02 m^3; 3·0)

2.1 A stream flows in a rectangular channel 10 m wide. The depth of flow is 0·2 m and the mean channel velocity is 0·5 m s^{-1}. The temperature of the water is 10°C. Calculate the Reynolds number of the flow.

(Answer: 294 000)

2.2 Overland flow is taking place on a bare hillside at a shear velocity of 1 cm s^{-1}. The ground surface is covered by sandy soil with a particle size of 1 mm. If the viscosity of the water is $1{\cdot}4 \times 10^{-2}$ cm^2 s^{-1}, determine whether the flow corresponds to a hydrodynamically rough or smooth boundary flow by calculating the thickness of the laminar sublayer.

(Answer: smooth; 1·624 mm)

2.3 A wide stream channel (sufficiently wide that the hydraulic radius may be taken equal to the depth) is littered with bed material of 1 cm diameter and a packing coefficient equal to 0·75. The downstream bed gradient is 5°. Assuming that the specific gravity of the bed material is 2·6 and the angle of interlock among the particles is 35°, calculate the depth of flow in the channel necessary to initiate bed load movement, using White's approach adapted to a sloping surface.

(Answer: 4·2 cm)

2.4 A flat horizontal desert surface is mantled by debris 0·09 cm in diameter. Calculate the critical shear velocity necessary to initiate wind erosion of this material.

(Answer: 43 cm s^{-1})

3.1 A specimen of sandy soil is subjected to a direct shear test under an effective normal stress of 2·00 kg cm^{-2}; at failure the shear stress in the specimen attained a value of 1·14 kg cm^{-2}. Calculate (a) the ϕ' angle and (b) the shear stress necessary for failure under a normal stress of 3·50 kg cm^{-2}, assuming the soil to be cohesionless.

(Answer: 30°; 1·99 kg cm^{-2})

3.2 A triaxial test is undertaken on a saturated clay specimen. The value of the all-round cell pressure σ_3 is 2·0 bars; at failure the vertical stress σ_1 attains a value of 4·8 bars, at which time the pore pressure is 1·8 bars. Failure is observed to occur on a plane inclined at an angle of 57° with the horizontal. Calculate the shear stress, normal stress, and shear strength on (a) the failure plane and (b) the plane of maximum shear stress.

(Answer: 1·27 bars; 1·03 bars; 1·27 bars; 1·40 bars; 1·60 bars; 1·51 bars)

3.3 The major and minor principal stresses at a point in a stressed body are 4882 kg m^{-2} and 1465 kg m^{-2}, respectively. Determine the normal and shearing stresses on a plane that passes through the point and makes an angle of 25° with the major principal plane. Also determine the resultant of the normal and shear stresses, and find the angle which the resultant makes with the plane.

(Answer: 4267 kg m^{-2}; 1309 kg m^{-2}; 4461 kg m^{-2}; 17°)

4.1 Slopes in a sandy deposit are subject to instability at times when the material becomes fully saturated and horizontal groundwater flow occurs. The deposit is homogeneous and isotropic; it has a bulk unit weight of 2000 kg m^{-3}, a friction angle of 35°, and is completely cohesionless. Calculate the angle of slope at which the deposit just becomes stable.

(Answer: 17·5°)

4.2 Stream erosion is cutting a vertical-sided gully into a clay mass. The clay has the following engineering properties: $\gamma = 1762$ kg m^{-3}; $c = 1465$ kg m^{-2}; and $\phi = 25°$. Using the Culmann method of stability analysis calculate the maximum gully depth before the side walls collapse, assuming that the soil properties do not change over time and tension cracks do not develop prior to failure. Compare your answer with the critical height as determined from Figure 4.9.

(Answer: 5·3 m; 5·1 m)

4.3 A deep gorge (500 m) has been cut into a massive limestone block which, when tested intact, shows unconfined compressive strength values of about 500 kg cm^{-2}. The unit weight of the rock is 2500 kg m^{-3}. Theoretical studies estimate that the strength of the rock is decreasing, due to pressure release and stress concentration effects, at a rate given by

$$\lg y = 30x$$

where y is the strength loss in kg cm^{-2} and x is the time elapsed since the origin of the gorge, in million years. Calculate when, relative to the cutting of the gorge, instability is first likely to occur; use the Culmann stability method, and assume that tension cracks open up prior to failure.

(Answer: 86000 y)

5.1 A temperate glacier, averaging 100 m in thickness and with a surface slope parallel to the bed slope of 6°, rests on a bed with a roughness (Weertman) of 4 and a constant protuberance size (L) equal to 1 cm. Calculate the theoretical velocity of basal sliding using, firstly, Weertman's equation for pressure-melting (5.13), and, secondly, the equation for accelerated plastic creep (5.18). Ignore any resistance along the sides of the glacier. Use the following values:

$H = 80$ cal g^{-1}
$K = 5 \times 10^{-3}$ cal °C^{-1} cm^{-1} s^{-1}
$C = 7{\cdot}4 \times 10^{-9}$
$\rho_i = 1$ g cm^{-3}
$A' = 0{\cdot}017$ bar^{-4} y^{-1}
$n = 4$

(Answer: 2·3 m y^{-1}; 6·6 m y^{-1})

5.2 Using the same data as in the previous question calculate the theoretical surface velocity relative to the basal velocity.

(Answer: 59 m y^{-1})

5.3 A granite block, 1 m^3 in size, rests on the flat horizontal bed of a glacier, with its downvalley face normal to the direction of ice flow, abutting against a protuberance from the glacier bed. The shear strength of the bedrock, when intact, is 160 kg cm^{-2}. Calculate the maximum size of protuberance, in terms of plan area, which could be planed off by the force of the block. Assume that: ice adheres to the top and sides of the block; a gap exists between the downvalley face of the block and the ice; the ice has a yield strength in compression of 2 kg cm^{-2}, and a friction coefficient of zero; and there is no frictional resistance between the underside of the block and the glacier bed. Repeat the calculation making the more realistic assumption that a friction coefficient of $\mu = 0{\cdot}7$ exists at the base of the block and, in addition, that there are no shear forces due to ice contact on any face of the block. Assume a value for the specific gravity of the block equal to 2·6.

(Answer: 313 cm^2; 25 cm^2)

5.4 Consider (in relation to Figure 5.16) a cirque glacier with surface slope equal to 20° and a bed contact corresponding to $a = 1$. Calculate the ice thickness H necessary to drive the glacier forward in a rotational motion, using the data of Figure 4.9 and a yield strength for ice in shear equal to 1 kg cm^{-2}. Calculate the extra thickness of ice necessary to achieve abrasion of the underlying rock surface, assuming that the average friction coefficient on the rotational shear surface is 0·1.

(Answer: 66 m; 64 m)

Index